ALSO BY NEIL SHUBIN

The Universe Within

Your Inner Fish

SOME ASSEMBLY REQUIRED

SOME ASSEMBLY REQUIRED

Decoding Four Billion Years of Life, from Ancient Fossils to DNA

Neil Shubin

PANTHEON BOOKS, NEW YORK

Published in the United States by
Pantheon Books, a division of Penguin Random
House LLC, New York, and distributed in Canada
by Penguin Random House Canada Limited, Toronto.

Pantheon Books and colophon are registered
trademarks of Penguin Random House LLC.

Library of Congress Cataloging-in-Publication Data
Name: Shubin, Neil, author.
Title: Some assembly required: decoding four billion years
of life, from ancient fossils to DNA / Neil Shubin.
Identifiers: LCCN 2019022792. ISBN 9781101871331 (hardcover).
ISBN 9781101871348 (ebook)
Subjects: LCSH: Life—Origin. Paleontology. Evolution (Biology)
Classification: LCC QH325 .S544 2020 | DDC 576.8—dc23
LC record available at lccn.loc.gov/2019022792

www.pantheonbooks.com

Original illustrations by Kalliopi Monoyios

Jacket images from *Die Vergleichende Osteologie*,
Bonn, 1821 (details). AAA Collection/Historic Images/Alamy
Jacket design by Perry De La Vega

Printed in the United States of America
First Edition
2 4 6 8 9 7 5 3 1

To the memory of my parents,
Seymour and Gloria Shubin

CONTENTS

Prologue ix

1. Five Words 3
2. Embryonic Ideas 28
3. Maestro in the Genome 60
4. Beautiful Monsters 92
5. Copycats 124
6. Our Inner Battlefield 146
7. Loaded Dice 168
8. Mergers and Acquisitions 193

Epilogue 215

Further Reading and Notes 219

Acknowledgments 251

Illustration Credits 255

Index 257

PROLOGUE

Decades spent cracking rocks have changed the way I see living things. If you know how to look, scientific research becomes a global treasure hunt for fossils of fish with arms, snakes with legs, and apes that can walk upright, all ancient creatures that tell about important moments in the history of life. In *Your Inner Fish*, I described how planning and luck led my colleagues and me to find *Tiktaalik roseae* in the High Arctic of Canada: a fish with a neck, elbows, and wrists. This creature bridged the gap between life in water and life on land, to reveal the important moment when our distant ancestors were fish. For almost two centuries, discoveries like these have told us how evolution happens, how bodies are built, and how they came into being. But paleontology has arrived at an important moment of change, one that coincided with the start of my career almost four decades ago.

Growing up on *National Geographic* magazine and television documentaries, I knew from a relatively early age that I wanted to join expeditions to discover fossils. This interest led me to graduate school at Harvard University, where I ended up leading my first fossil-hunting trips in the mid-1980s. Lacking the ability to launch excursions to exotic locales, I explored the rocks along roadsides south of Cambridge, Massachusetts. Returning from the field after one of these trips, I found a pile of journal articles atop my desk. This stack of papers was my introduction to how the world of paleontology was about to change dramatically.

A fellow graduate student found articles in the library that described how a number of laboratories had discovered DNA that helps build animal bodies, revealing genes that work to make the heads, wings, and antennae of flies. That fact alone was incredible, but there was more: versions of the same genes were making the bodies of fish, mice, and people. The pictures in these papers held glimmers of a new science, one that could explain how animals are assembled in the embryo and how they evolved over millions of years.

Experiments with DNA promised to answer questions that were formerly the sole purview of fossil hunters. Moreover, an understanding of DNA could get to the genetic mechanism that drove the changes I was seeking to explain among ancient rocks.

Like fossil species in our past, I was going to have to evolve or go extinct. If extinction for a scientist is irrelevance, then a deep dive into genetics, developmental biology, and the world of DNA would keep me part of the intellectual action. Ever since those first journal articles, I have run a kind of split-brain laboratory, spending summers in the field looking for fossils and working the rest of the year with embryos and DNA. Both approaches can be deployed in the service of answering a single question: How do large changes in the history of life come about?

For the past two decades, technological advances have arrived at a dizzying pace. Genome sequencers are now so powerful that the Human Genome Project, which took over a decade and cost billions of dollars, could now be completed in an afternoon for under one thousand dollars. And sequencing is only one example: computing power and imaging technologies allow us to peer inside embryos and even to see molecules at work in cells. DNA technology has become so powerful that animals as diverse as frogs and monkeys can now be readily cloned, and mice can be engineered with the genes of humans or flies inside. The DNA of almost any animal can now be edited, giving us the power to remove and rewrite the genetic code that builds bodies of almost every species of animal and plant. We can ask, at the level of DNA, what combination of genes makes a frog different from a trout, a chimpanzee, or a human?

This revolution has brought us to a remarkable moment. Rocks and fossils, when coupled with DNA technology, have the power to probe some of the classic questions that Darwin and his contemporaries struggled with. New experiments reveal a multibillion-year history filled with cooperation, repurposing, competition, theft, and war. And that is just what happens inside DNA itself. With viruses continually infecting it, and its own parts at war with one another, the genome within each animal cell roils as it does its work in generation after generation. The outcome of this dynamism has been new organs and tissues, biological innovations that have changed the world.

Once life emerged, the entire planet was a microbial zoo for billions of years. About a billion years ago, single-celled microbes gave rise to creatures with bodies. A few hundred million years more saw the origin of everything from jellyfish to people. Since that time, creatures have evolved to swim, fly, and think as each invention presaged the next. Birds use wings and feathers to fly. Animals that live on land have lungs and limbs. The list goes on. From simple ancestors, animals have evolved to live at the bottom of the ocean, inhabit barren deserts, thrive on the tops of the highest mountains, and even walk on the moon.

The great transformations in the history of life have brought about wholesale shifts in how animals live and how their bodies are organized. The evolution of fish to land-living creatures, the origin of birds, and the beginnings of bodies themselves from single-celled creatures—these are but a small number of revolutions in the history of life. And the science that probes them is full of surprises. If you think feathers arose to help animals fly, or lungs and legs to help animals walk on land, you'd be in good company. You'd also be entirely wrong.

Advances in this science can help answer some of the basic questions of our existence: Is our presence on this planet the result of chance? Or was the history that brought us here inevitable in some way?

The history of life has been a long, strange, and wondrous trip of trial and error, chance and inevitability, detours, revolution, and invention. That path, and the way we have come to know it, is the story of this book.

SOME ASSEMBLY REQUIRED

1

Five Words

SOME PEOPLE FIND THE SUBJECT of their life's work in a laboratory or in the field. I found mine in a single projected slide.

Early in my graduate student days, I took a class taught by a senior scientist on the greatest hits in the history of life. It was a whirlwind course, a form of speed dating with big puzzles in evolution. Fodder for each week's discussion was a different evolutionary transformation. In one of the initial sessions, the professor displayed a simple cartoon that showed what we knew back then, in 1986, about the transition from fish to land-living animals. At the top of the sketch was a fish and at the bottom was an early fossil amphibian. An arrow pointed from the fish to the amphibian. It was the arrow, not the fish, that caught my eye. I looked at that figure and scratched my head. Fish walking on land: How could that ever happen? This seemed like a first-class scientific puzzle on which to hang my shingle. It was love at first sight. Thus began four decades of expeditions to both poles, and several continents, in the hunt for fossils to show how this event transpired.

Yet when I tried to explain my quest to relatives and friends, I was often met with pained glances and polite questions. Transforming a fish into a land-living animal meant developing a new kind of skeleton, one with limbs for walking rather than fins for swimming. Moreover, a new way of breathing, using lungs rather than gills, had to arise. So, too, feeding and reproducing had to change—eating and laying eggs in water is entirely different from what happens on land. Virtually every system in the body would have to transform simultaneously. What good would it be to have limbs for walking on land if the animal couldn't breathe, feed, or reproduce? Living on land requires not just a single invention but the interplay of hundreds of them. The same difficulty holds for each of the thousands of other transitions in the history of life, from the origins of flight and bipedal walking to the origins of bodies and life itself. My quest seemed doomed from the start.

The solution to this dilemma is embedded in a famous quote from the playwright Lillian Hellman. In describing her life—from being blacklisted by the House Un-American Activities Committee during the 1950s to her hard-living ways—she once said, "Nothing, of course, begins at the time you think it did." With that phrase, she unintentionally described one of the most powerful concepts in life's history, one that explains the origin of most every organ, tissue, and bit of DNA in all creatures on Planet Earth.

The seeds for this idea in biology began as a consequence of the work of one of the most self-destructive figures in all of science, who, true to form, changed the field by being wrong.

To grasp the meaning of recent discoveries in the genome, we need to turn to an earlier age of exploration. Victorian England was a crucible for enduring ideas and discoveries. There is something poetic to the notion that knowing how DNA works in the history of life relies on ideas developed during an age when people didn't know that genes even existed.

St. George Jackson Mivart (1827–1900) was born to zealously evangelical parents in London. His father had risen from being a butler to owning one of the city's major hotels. Mivart Senior's position gave his son the chance to achieve the social standing of a gentleman and accorded him the privilege of entrée into the career of his choice. Like his contemporary Charles Darwin, Mivart was born with a passion for nature. As a child, he collected insects, plants, and minerals, often making copious field notes and devising classification schemes. Mivart seemed destined for a career in natural history.

Then the dominant theme of his personal life—struggle with authority—intervened. In his preteens, Mivart became increasingly uncomfortable with his family's Anglican faith. To the great consternation of his parents, he converted to Roman Catholicism. This move, bold for a sixteen-year-old, had unforeseen consequences. Mivart's newfound allegiance to the Catholic Church meant that he couldn't attend Oxford or Cambridge, because entrance to English universities was closed to Catholics at that time. Unable to matriculate to any program in natural history, he took the only remaining option—studying law at the Inns of Court, where one's choice of religion was not an obstacle. Mivart became a lawyer.

It is not clear if Mivart ever practiced law, but natural history remained his passion. Using his status as a gentleman, he entered scientific high society, where he developed relationships with key figures of the day, most notably Thomas Henry Huxley (1825–95), who was soon to become a prominent defender of Darwin's ideas in the public sphere. Huxley was an accomplished comparative anatomist in his own right and had assembled a cadre of keen apprentices. Mivart became close to the great man, working in his lab, even taking part in Huxley family gatherings. Under Huxley's tutelage, Mivart produced seminal, albeit mostly descriptive, works in primate comparative anatomy. These detailed accounts of the skeleton remain useful today. By the time Darwin published his first edition of *On the Origin of Species* in 1859, Mivart considered himself a supporter of Darwin's new idea, likely a by-product of being enveloped by Huxley's fervor.

But, as had happened with the Anglican faith of his youth, Mivart developed strong doubts about Darwin's ideas and intellectual objections to the Darwinian idea of gradual change. He began to voice his notions in public, first meekly, then with greater force. Marshaling evidence in support of his dissenting view, he composed a response to *On the Origin of Species*. If he had any remaining friends among his old pals in the natural his-

St. George Jackson Mivart, who managed to offend every side in the evolution debate

tory world, he lost them with his single-word variant of Darwin's title: *On the Genesis of Species*.

Mivart then started giving the Catholic Church a hard time too. He wrote in church periodicals that virgin birth and the infallibility of church doctrine were as implausible as Darwin's ideas. With the publication of *On the Genesis of Species*, Mivart was virtually excommunicated from science. His writings led the Catholic Church to formally excommunicate him six weeks before his death in 1900.

Mivart's challenge to Darwin offers a window into the intellectual knife fights of Victorian England and articulates a stumbling block that many people continue to have with Darwin. Mivart opened his attack by referring to himself in the third person, using language intended to establish his credibility as open-minded: "He was not originally disposed to reject Darwin's fascinating theory."

Mivart begins making his case with a substantial chapter outlining what he saw as Darwin's fatal flaw, calling it "the incompetency of natural selection to account for the incipient stages of useful structures." The title is a mouthful, but it encapsulates a crucial issue: Darwin envisioned evolution as consisting of innumerable intermediate stages from one species to another. For evolution to work, each of these intermediate stages had to be adaptive and increase an individual's ability to thrive. To Mivart, intermediate stages often didn't appear plausible. Take the origin of flight, for example. What possible use could an early stage in the origin of wings have? The late paleontologist Stephen Jay Gould called this issue the "2% of a wing problem": a tiny incipient wing in a bird ancestor would appear to have no utility at all. At some point it might be big enough to help an animal glide, but a tiny wing couldn't be used for any type of powered flight.

Mivart offered one case after another in which intermediate

stages seemed implausible. Flatfish have two eyes on one side of the body, giraffes have long necks, some whales have baleen, various insects mimic tree bark, and on and on. What use could tiny fractional displacements of the eyes, elongations of necks, or subtle variations in coloration have? How about a jaw with only a sliver of baleen to feed an entire whale? Evolution, it would appear, consisted of innumerable dead ends between the endpoints of any major transition.

Mivart was one of the first scientists to call attention to the observation that major transitions in evolution do not involve a single organ changing; rather, whole suites of features across the body have to change in concert. What was the use of evolving limbs to walk on land if a creature didn't have lungs to breathe air? Or, as another example, consider the origin of bird flight. Powered flight requires many different inventions—wings, feathers, hollow bones, high metabolisms. It would be useless for a creature with bones as clunky as an elephant's or a metabolism as slow as a salamander's to evolve wings. If entire bodies have to change for any great transformation, and many features need to change simultaneously, then how could major transitions happen gradually?

In the century and a half since the publication of Mivart's ideas, they have been a touchstone for many critiques of evolution. At the time, however, they also served as a catalyst for one of Darwin's great ideas.

Darwin saw in Mivart a truly important critic. He published the first edition of *On the Origin of Species* in 1859; Mivart's tome appeared in 1871. For the sixth, definitive edition of *On the Origin of Species*, published in 1872, Darwin added a new chapter to respond to his critics, Mivart chief among them.

True to the conventions of Victorian debate, Darwin opened by saying, "A distinguished zoologist, Mr. St. George Mivart, has recently collected all the objections which have ever been advanced by myself and others against the theory of natural selection, as propounded by Mr. Wallace and myself, and has illustrated them with admirable art and force." He continued: "When thus marshaled, they make a formidable array."

Then he silenced Mivart's critique with a single phrase, followed by copious examples of his own. "All Mr. Mivart's objections will be, or have been, considered in the present volume. The one new point which appears to have struck many readers is, 'That natural selection is incompetent to account for the incipient stages of useful structures.' This subject is intimately connected with that of the gradation of the characters, often accompanied *by a change of function.*"

It is hard to overestimate how deeply important those last five words have been to science. They contain the seeds for a new way of seeing major transitions in the history of life.

How is this possible? As usual, fish provide insights.

Breath of Fresh Air

When Napoleon Bonaparte invaded Egypt in 1798, he brought more than ships, soldiers, and weapons with his army. Seeing himself as a scientist, he wanted to transform Egypt by helping it control the Nile, improve its standard of living, and understand its cultural and natural history. His team included some of France's leading engineers and scientists. Among them was Étienne Geoffroy Saint-Hilaire (1772–1844).

Saint-Hilaire, at twenty-six, was a scientific prodigy. Already chair of zoology at the Museum of Natural History in Paris, he was destined to become one of the greatest anatomists of all time. Even in his twenties, he distinguished himself with his ana-

Étienne Geoffroy Saint-Hilaire, scientific prodigy

tomical descriptions of mammals and fish. In Napoleon's retinue he had the exhilarating task of dissecting, analyzing, and naming many of the species Napoleon's teams were finding in the wadis, oases, and rivers of Egypt. One of them was a fish that the head of the Paris museum later said justified Napoleon's entire Egyptian excursion. Of course, Jean-François Champollion, who deciphered Egyptian hieroglyphics using the Rosetta Stone, likely took exception to that description.

With its scales, fins, and tail, the creature looked like a standard fish on the outside. Anatomical descriptions in Saint-Hilaire's day entailed intricate dissections, frequently with a team of artists on hand to capture every important detail in beautiful, often colored lithographs. The top of the skull had two holes in the rear, close to the shoulder. That was strange enough, but the real surprise was in the esophagus. Normally, tracing the esophagus in a fish dissection is a pretty unremarkable affair, as it is a simple tube that leads from the mouth to the stomach. But this one was different. It had an air sac on either side.

This kind of sac was known to science at the time. Swim bladders had been described in a number of different fish; even Goethe, the German poet and philosopher, once remarked on them. Present in both oceanic and freshwater species, these sacs fill with air and then deflate, offering neutral buoyancy as a fish navigates different depths of water. Like a submarine that expels air following the call to "dive, dive, dive," the swim bladder's air concentration changes, helping the animal move about at varying depths and water pressures.

More dissection revealed the real surprise: these air sacs were connected to the esophagus via a small duct. That little duct, a tiny connection from the air sac to the esophagus, had a large impact on Saint-Hilaire's thinking.

Watching these fish in the wild only confirmed what Saint-Hilaire inferred from their anatomy. They gulped air, pulling it in through the holes in the back of their heads. They even exhibited a form of synchronized air sucking, with large cohorts of them snorting in unison. Groups of these snuffling fish, known as bichirs, would often make other sounds, such as thumps or moans, with the swallowed air, presumably to find mates.

The fish did something else unexpected. They breathed air. The sacs were filled with blood vessels, showing that the fish were using this system to get oxygen into their bloodstreams. And, more important, they breathed through the holes at the top of their heads, filling the sacs with air while their bodies remained in the water.

Here was a fish that had both gills and an organ that allowed it to breathe air. Needless to say, this fish became a cause célèbre.

A few decades after the Egyptian discovery, an Austrian team was sent on an expedition to explore the Amazon in celebration of the marriage of an Austrian princess. The team collected insects, frogs, and plants: new species to name in honor of the royal family. Among the discoveries was a new fish that, like any fish, had both gills and fins. But inside it also had unmistakable vascular plumbing: not a simple air sac, but an organ loaded with the lobes, blood supply, and tissues characteristic of true human-like lungs. Here was a creature that bridged two great forms of life: fish and amphibians. To capture the confusion, the explorers gave it the name *Lepidosiren paradoxa*—Latin for "paradoxically scaled salamander."

Call them what you will—fish, amphibian, or something in between—these creatures had fins and gills to live in water but also lungs to breathe air. And they weren't just one-offs. In 1860 still another fish with lungs was discovered in Queensland, Australia. This fish also had a very distinctive set of teeth. Shaped

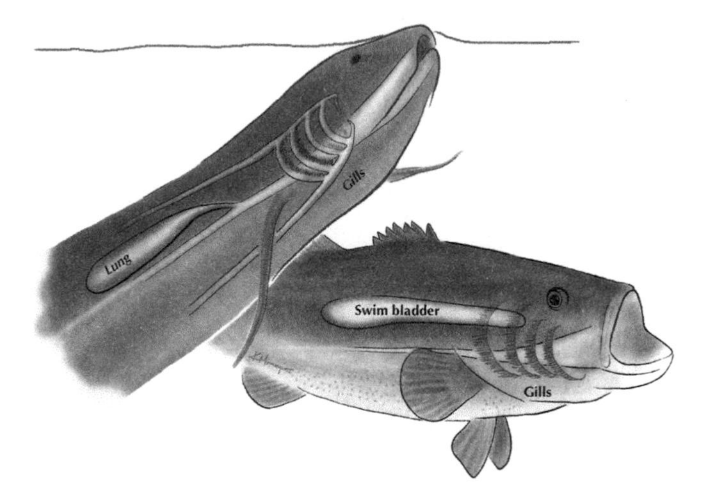

Lungfish have both lungs and gills. They use lungs like ours to breathe air when the oxygen content of the water doesn't meet their needs. Other fish have swim bladders that aid in buoyancy.

like a flat cookie cutter, such teeth were known from the fossil record from a species that was long extinct—an animal named *Ceratodus* found in rocks over 200 million years old. The implication was clear: lunged, air-breathing fish were global and had been living on Planet Earth for hundreds of millions of years.

An aberrant observation can be a game changer for how we see the world. Fish lungs and swim bladders spawned a generation of scientists interested in exploring the history of life by looking both at fossils and at living creatures. Fossils show what life looked like in the distant past, and living creatures reveal how anatomical structures work, as well as how organs develop from egg to adult. As we'll see, this is a powerful approach.

Linking studies of fossils and embryos was a fruitful area of inquiry for the natural scientists who followed Darwin. Bashford Dean (1867–1928) had an unusual distinction in academic circles—he is the only person ever to hold a curatorship at both the Metropolitan Museum of Art and, directly across Central Park, the American Museum of Natural History. He had two passions in life, fossil fish and battle armor. He founded the armor collection and displays at the Met, and he did the same for the fish collection at the Museum of Natural History. Befitting a person with such interests, he was a quirky individual. He designed his own armor and even took to wearing it on the streets of Manhattan.

When he wasn't donning medieval faulds, Bashford Dean was studying ancient fish. Somewhere locked inside the embryo's transformation from egg to adult, he believed, were answers to the mysteries of history and the mechanism of current fish's descent from ancestral species. Comparing fish embryos with fossils and reviewing the work in anatomy labs at the time, Dean saw that lungs and swim bladders look essentially the same dur-

Bashford Dean, a curator at the Metropolitan Museum of Art and at the American Museum of Natural History, loved both battle armor and fish.

ing development. Both organs bud from the gut tube and both form air sacs. The major difference is that swim bladders develop on the top of the tube, near the spine, while lungs bud from the bottom, or belly side. Using these insights, Dean argued that swim bladders and lungs were different versions of the same organ, formed by the same developmental process. Indeed, some kind of air sac is present in virtually all fish but sharks. Like many ideas in science, Dean's comparison has a long history. Its antecedents can be seen in the work of nineteenth-century German anatomists.

But what do air sacs say about Mivart's critique and Darwin's response?

A surprising number of fish can breathe air for extended periods of time. The six-inch-long mudskipper can walk and live on the mud for over twenty-four hours. The aptly named climbing perch can wiggle from pond to pond as needed, sometimes even climbing branches and stepping over twigs in the process. But that perch is only a single species. Hundreds of species can gulp air when the concentration of oxygen in the water they inhabit declines. How do these fish do it?

Some, like the mudskipper, absorb oxygen through their skin. Others have a special gas-exchange organ above their gills. Some catfish and other species absorb oxygen through their guts, gulping air like food, only to use it to breathe. And a number of fish have paired lungs that look like our own. Lungfish live in water and breathe with their gills most of the time, but when the oxygen content of their stream is not sufficient to support their metabolism, they will push to the surface and gulp air into their lungs. Air breathing is not some crazy exception in an oddball fish—it is the common state of affairs.

Recently, researchers at Cornell University revisited the comparison of swim bladders to lungs, using new genetic techniques. Their question: What genes help build fish swim bladders during development? In looking at the catalog of genes that are active in fish embryos, they found something that would have pleased both Dean and Darwin. The genes that are used to build swim bladders in fish are the same ones used to make lungs in both fish and people. Having an air sac is common to virtually all fish; some use them as lungs, while others use them as buoyancy devices.

Here is where Darwin's answer to Mivart becomes so prescient. DNA clearly shows that lungfish, Saint-Hilaire's bichirs, and other fish with lungs are the closest living fish relatives to land-living creatures. Lungs aren't some invention that abruptly came about as creatures evolved to walk. Fish were breathing air with lungs well before animals ever stepped onto terra firma. The invasion of land by descendants of fish did not originate a new organ—it changed the function of an organ that already existed. Moreover, virtually all fish have some kind of air sac, whether lung or swim bladder. Air sacs shifted from being used for a life in water to later enabling creatures to live and breathe on land. The change did not involve the origin of a new organ; instead the transformation was, as Darwin said more generally, "accompanied *by a change of function.*"

Causing a Flap

The target of Mivart's complaint against Darwin hadn't been fish or amphibians but birds. At the time, the origin of flight was a colossal mystery. In the first edition of *On the Origin of Species* in 1859, Darwin made very specific predictions. If his theory of a common ancestry for life on Earth was true, there should be intermediates in the fossil record, ones that represent transitions between different forms of life. At the time, none were known, let alone any that linked flying birds to creatures that dwelled on the ground.

Darwin did not have to wait long, however. In 1861, workers at a limestone quarry in Germany discovered a remarkable fossil. The quarry's fine-grained limestone made it an ideal stone for the slabs used in lithography, the printing process of the day. The limestone was formed in a very gentle lake environment, meaning that whatever was captured inside it was relatively undisturbed. These rocks can be nearly perfect for preserving fossils.

This slab held a curious impression, capturing something long and pinnate. It looked like a perfectly formed feather. But why there would be a feather in these rocks was a mystery.

The limestone that held the strange impression dated from the Jurassic Age. Decades before this discovery, the German aristocrat and naturalist Alexander von Humboldt (1769–1859) had noticed distinctive limestone in the Jura Mountains, bordering France and Switzerland. This limestone formed a layer that extended for miles. Von Humboldt named it Jurassic for its distinctive features, suggesting that it might date to a special age in the history of the Earth. Soon afterward other scientists noticed that the Jurassic layer is often filled with fossils, such as large coiled, shelled creatures known as ammonites. Similar fossils were found around the world, leading researchers to recognize the Jurassic as a distinctive age more globally, not particular to France and Switzerland.

Then, in the early 1800s, large teeth and jaws were found in Jurassic rocks in England. Similar discoveries started to crop up everywhere. It soon became clear that the Jurassic had been the era not only of coiled, shelled creatures but of dinosaurs. The feather impression revealed even more. Were birds flying above the dinosaurs on land during the Jurassic?

An isolated fossil of a feather was tantalizing. Perhaps it was attached to a Jurassic bird? Or maybe some unknown kinds of creatures also had feathers? That hypothesis could not be ruled out.

A few years after the discovery of the feather in 1861, a farmer traded a fossil in exchange for medical services. This fossil came from the same limestone as the isolated feather. The doctor who bought it was a trained anatomist who had a passion for fossils. Consequently, he knew at first glance that this was no ordinary slab of limestone. The fossil inside had feather impressions that covered the body and tail, and they were attached to a nearly complete skeleton with hollow bones and wings. Knowing the specimen's value, the doctor opened up a bidding war among museums for it, eventually extracting 750 pounds from the British Museum.

Over the next fifteen years, more specimens turned up. In the mid-1870s a farmer named Jakob Niemeyer traded a fossil to a quarry owner for the price of a cow. The quarry owner, knowing the renown of the physician who had parlayed the previous specimen to London, sold the fossil to the same physician in 1881. This skeleton fetched a thousand pounds from the Museum of Natural History of Berlin. As of today, a total of seven specimens have been discovered.

The feather-covered creature, dubbed *Archaeopteryx*, had a curious mix of features. Like a bird, it had wings replete with feathers and hollow bones. But unlike any known bird, it had teeth like a carnivore, a flat breastbone, and three sharp claws on the bones at the tips of its wings.

This discovery couldn't have happened at a better time for Darwin's theory. When Thomas Henry Huxley examined the teeth, limbs, and claws of *Archaeopteryx*, he saw a deep resemblance between *Archaeopteryx* and reptiles. He compared *Archaeopteryx* to another creature from Jurassic limestone, a small dinosaur known as *Compsognathus*. The two creatures were of the same size and had a similar skeleton except for feathers. Huxley proclaimed *Archaeopteryx* to be proof of Darwin's theory—it was an intermediate between reptiles and birds. Darwin even made a reference to *Archaeopteryx* in his fourth edition of *On the Origin of Species:* "Hardly any recent discovery shows more forcibly than this how little we as yet know of the former inhabitants of the world."

Comparisons such as Huxley's ignited a wide-ranging controversy. If *Archaeopteryx* was evidence that birds were related to reptiles, which reptiles were their ancestors? There were several obvious candidates, each with its own defenders. Some proposed that the long tail of *Archaeopteryx* and form of its skull revealed that the ancestors of birds were small, carnivorous, lizard-like creatures. Others compared birds to another group of flying reptiles from the Jurassic, the pterosaurs. The difficulty with this theory was that while pterosaurs had wings and flew, the bones that formed their wings are very different from those of birds. Pterosaur wings are supported by an elongated fourth digit, while bird wings are supported both by feathers and by a combination of digits. Still others were impressed by Huxley's comparison of *Archaeopteryx* and the small dinosaur.

The idea that the ancestor of birds was some kind of dinosaur gained prominent detractors over the years, each relying on different arguments. One researcher claimed to find a fatal flaw in birds' dinosaurian ancestry: birds have clavicles whereas dinosaurs, unlike all other reptiles, do not. Other researchers saw dinosaurs and birds as completely different in lifestyle and metabolism, so much so that dinosaurs could never be seen as bird ancestors. Dinosaurs were, with few exceptions, large slow-moving beasts, not very similar to highly active small birds. *Archaeopteryx*, to many, was just a bird and did not say much about the transition. The struggle continued, largely because Mivart's essential criticism remained: How could feathers and all other specialized features of birds, including those of *Archaeopteryx*, have arisen?

The idea that dinosaurs were massive and lumbering beasts has a long history. So does the demise of this view, which began with the work of an eclectic scientist who, like Bashford Dean, loved to don military costumes.

Franz Nopcsa von Felső-Szilvás (1877–1933), known as Baron Nopcsa of Săcel, was a man of intense passions and great intellect. At eighteen, he discovered some bones on his family's estate in Transylvania. After teaching himself anatomy, in 1897 he published a formal scientific description of them as a large dinosaur. Nopcsa went on to write a seven-hundred-page tome on the geology of Albania, as well as dozens of scientific papers in multiple languages. He served as a spy for Austria and worked to organize Albanians' resistance to the Turks to gain their freedom. The baron's real dream was to assume the throne of Albania. Sadly, his life ended when, after racking up large debts, he shot his lover, then turned the gun on himself.

After his encounter with the bones on his family land in 1895, Nopcsa amassed a large fossil collection and took to studying Transylvanian dinosaurs, both their bones and the trackways they left in stones preserved throughout eastern Europe. Looking at trackways preserved in the rocks, he saw traces of living, breathing creatures walking through muds. The markings in the mud showed that the animals that left them could clearly run fast. These animals were pushing hard against the ground,

Baron Nopcsa in Albanian uniform. Like Dean, he studied the deep history of evolutionary innovations and also relished sporting armor and military regalia.

and the distance between the footprints revealed that they were using a running gait. The implication was clear—far from being slow-moving beasts like elephants, dinosaurs were fast-running and active predators. Nopcsa took this idea even further: because running dinosaurs would need to be fast and light, they would make excellent precursors for birds. The need for speed, in his view, would have driven them to the air, and feathered wings would have aided the protobirds to flap their arms to increase speed and catch prey.

When he published his idea in 1923, Nopcsa suffered the fate that is a nightmare of most scientists: he was ignored. The long-dominant theory, by this time forcefully promulgated by the eminent Yale paleontologist O. C. Marsh, held that dinosaurs were large and slow-moving, and that birds arose from ancestors that were gliders. Powered flight presumably had its origins in tree-dwelling animals that used gliding to move from branch to branch. Over time flight evolved from these gliding ancestors. The intuitive appeal of this theory is seen in the diverse gliding animals that exist today, from frogs and snakes to squirrels and lemurs. As relatively fewer complex inventions are needed to become a glider rather than a flier, gliding seemed a logical first step in the origin of powered flight.

In the 1960s John Ostrom, then a junior scientist at Yale, was trying to understand how duck-billed dinosaurs had lived. These familiar denizens of the dinosaur halls of almost all major museums often have huge crests in their skulls that project away from their eponymous beaks. For years, museum displays depicted them as slow-moving plant eaters that moved on four legs, almost like reptilian elephants. But the more Ostrom looked at the bones, the less sense this interpretation made. First off, the front limbs were relatively short. Puny forelimbs with robust hind limbs would have made them strangely hunched for an animal that walked on four legs. Moreover, the crests and projections on the hind limbs suggested they had powerful muscles to move them. Taken together, these observations implied that duckbills had been mostly bipedal. Ostrom went even further: he saw duckbills not as lumbering beasts like elephants but as relatively active two-legged runners. Bipedal buffalos, he called them.

The Mivart-Darwin exchange from the 1800s gained new meaning when Ostrom took to the badlands of Wyoming in the 1960s. Like most paleontologists, Ostrom lived two existences: that of a buttoned-down scholar and teacher during the school year and a dusty, rough-and-tumble life on expedition in summertime. In August 1964 he was finishing an unremarkable expedition near the town of Bridger, Montana, by scoping about for sites for the next year's work. Ambling down the side of a bluff, he and an assistant were stopped in their tracks by something sticking out of the rocks. It would turn out to be a hand, about six inches long. "We both nearly rolled down the slope in our rush to the spot," Ostrom later said, describing the experience. The reason for the rush lay in what extended from the hand: sharp outsize claws, the likes of which they had not seen before.

As this was a last-day reconnaissance hike, they had no tools on them. Students of paleontology who read this paragraph should ignore what they did next. Breaking the prime directive of paleontological fieldwork in their excitement, they dug rapidly with their hands and penknives to expose more of the beast. Returning the next day with proper tools, they exposed a foot and some teeth. The teeth were those of a predator, with a sharp point and serrated edges. Two more years of digging led to the recovery of much of the skeleton.

Ostrom's dinosaur was the size of a large dog, but its bones were strangely light and hollow. The creature had a muscular tail and extremely powerful hind limbs with claws. The claws were set on joints, implying they could be used to shuck prey. Ostrom named the beast *Deinonychus* (Greek for "terrible claw"). In his later scientific monograph, buried in the normally standard dry prose of the form, he described *Deinonychus* as being "highly predaceous, extremely agile, and very active."

Deinonychus was only the beginning. Ostrom and those who followed him changed how we think of dinosaurs and, in the process, exposed the power of Darwin's response to Mivart. They looked at every bump, hole, and feature on reptile bones and compared them to the bones of fossil and living birds. They soon concluded that dinosaurs, particularly the bipedal ones, and birds shared many characteristics. These species, theropod dinosaurs, have suites of bird features, including hollow bones and relatively fast growth rates. They were likely very active animals with high metabolisms.

Although these dinosaurs had numerous similarities to birds,

Deinonychus, the "terrible clawed" dinosaur

they were missing one important feature: feathers. Feathers were seen as the sine qua non of being avian, associated with the success of birds and the origin of flight.

In 1997 the Society of Vertebrate Paleontology held its meeting at the American Museum of Natural History in New York. Most of us in attendance knew something strange was afoot. This international gathering is usually a pretty staid affair, with talks and posters punctuated by cocktail parties and social events. At the time, members of the society tended to fragment into cliques, mostly defined by the creatures they worked on. Mammal researchers would attend mammal presentations, fish paleontologists would go to fish talks, and so on. We would socialize, then go our separate ways for the scientific sessions.

But 1997 was different. There was a buzz in the air in every hall and in every clique: "Have you seen it?" "Is it for real?"

Chinese colleagues had shown up with pictures of a new beast that had been discovered by farmers in the province of Liaoning, just northeast of Beijing. With hollow bones, clawed hands and feet, and a long tail, it had all the characteristics of a *Deinonychus*-like dinosaur. But this fossil was exquisitely preserved. It was embedded in the fine grains characteristic of rocks that preserve impressions or fragments of fossilized soft tissues. And that was what the buzz was all about: surrounding the dinosaur were unmistakable feathers. Not full feathers, but very simple downy ones. This dinosaur had had a primitive feathered covering.

Ostrom was in attendance. I was a junior scientist at the time and remember seeing him at a coffee break between sessions, talking to one of the more senior paleontologists. He was crying. His thirty years of controversial work had been vindicated by a fossil. At the time, he was quoted as saying, "I literally got weak in the knees when I first saw photos. The apparent covering on this dinosaur is unlike anything we have seen anywhere in the world before." He was later to say, "I never expected to see anything like this in my lifetime."

The feathered dinosaurs we saw in New York in 1997 were the first of a tidal wave of new fossils discovered in these Chinese sites. In the following decades, roughly twelve species of feathered dinosaurs emerged from China, painting a picture of carnivorous dinosaurs with a range of coverings. The most primitive of the lot have feathers of a simple tubular shape. The dinosaurs most closely related to *Archaeopteryx* and birds, however, have true feathers with a central shaft and fibers extending outward. Feathers are not a highly specialized feature of birds; they are found in virtually all carnivorous dinosaurs.

Birds are distinguished by more than feathers: they have wishbones, wings, and specialized wristbones used for flight. A

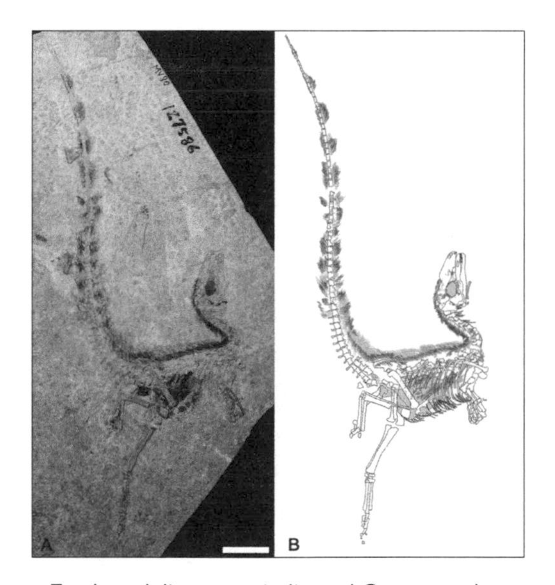

Feathered dinosaurs vindicated Ostrom and others who said that dinosaurs are the closest relatives of birds.

bird wing has the classic pattern of one bone, two bones, wristbones, and digits. Bird limbs only have three digits, not five, and the central one is elongated, serving as a point of attachment for feathers. Birds have fewer wristbones, including one that is shaped like a large crescent moon, the aptly named semilunate bone.

The more we look, the more we see that the anatomical inventions that birds use to fly, such as feathers, are not unique to them. Carnivorous dinosaurs get successively more birdlike over time. Primitive species have five-fingered limbs. Over tens of millions of years species lose digits until they are left with the bird pattern of three, including an enlarged central one that in birds serves as the base of the wing. Like birds, these dinosaurs lose wristbones and develop a semilunate bone, akin to the one that birds use in flapping flight. They even develop wishbones. None of these dinosaurs can fly, but all of them have some sort of feathers, from a simple downy covering in primitive forms to those with greater organization like *Archaeopteryx* and later dinosaurs. So what did feathers do in dinosaurs? Some paleontologists have proposed that they served as a kind of display to help them find mates. Others have suggested that primitive downy feathers served as a form of insulation, keeping the internal temperatures of the body warm. Perhaps feathers served in both roles. Whatever their function in dinosaurs, the origin of feathers is most definitely not related to flight.

Like lungs and limbs in the water-to-land transition, the inventions used for flight preceded the origin of flight. Hollow bones, fast growth rates, high metabolisms, winglike arms, wrists with hinges, and, of course, feathers originally arose in dinosaurs that were living on the ground, running fast to capture prey. The major change is not the development of new organs per se but the repurposing of old features for new uses and functions.

It has been common knowledge that feathers arose to help birds fly and lungs to enable animals to live on land. These notions are logical, obvious—and false. What's more, we've known this for over a century.

The not-so-hidden secret is that *biological innovations never come about during the great transition they are associated with.* Feathers did not arise during the evolution of flight, nor did lungs and limbs originate during the transition to land. What's more, these great revolutions in the history of life, and others like them, could never have happened otherwise. Major changes in the history of life didn't have to wait for the simultaneous origin of many inventions. Massive change came about by repurposing ancient structures for new uses. Innovations have antecedents that extend deep in time. Nothing ever begins when you think it does.

This is the story of revolution by evolution. Change in the history of life follows a twisted path, filled with detours, dead ends, and inventions that failed only because they arose at the wrong time. Darwin's five words, arguing that much of invention happens by a change in function of preexisting features, paved the way for our understanding of the origins of organs, proteins, even our DNA.

But the bodies of fish, dinosaurs, and people don't emerge fully formed at conception. They are built anew in each generation by a recipe transmitted from parent to offspring. The mother of invention lies inside these recipes and in how, as Darwin foresaw, they could arise in one context and, as we'll see, become repurposed in another.

2

Embryonic Ideas

CARL LINNAEUS (1707–78), the father of modern taxonomy, studied hundreds of plants and animals during his lifetime. His scientific classifications left little room for sentiment—except in one case. Of the thousands of animals Linnaeus investigated, he reserved one in particular for scorn and derision. Kids know salamanders and newts as gentle big-eyed creatures with large heads, four limbs, and long tails. But Linnaeus, for some unknown reason, thought them such "foul and loathsome animals" that he proclaimed it fortunate that "the creator has not exerted his powers to make many of them."

If Linnaeus saw salamanders as the nadir of creation, others claimed them to be elemental, almost magical, creatures. Philosophers from Pliny the Elder to Saint Augustine envisioned newts and salamanders as creatures born from lava, inferno, or flame. To Augustine, salamanders were physical evidence for the reality of damnation in fire. Augustine's idea derives from the claim that salamanders were resistant to flames or able to spring forth from bonfires. These superpowers may have reflected their biology. As aquarists and aficionados know, some salamander species have an affinity for the rotting undersides of logs. These wet habitats were likely hidden from those who in Augustine's day collected logs for firewood. When they ignited salamander-infused logs, they would have had some wiggly surprises that undoubtedly led to awe-inspired speculation about devilry.

While there are relatively few salamander species in the world, perhaps five hundred by some recent estimates, their relevance to the human condition lies well beyond visceral hatred, thoughts of damnation, and life emerging from fire. They have been a catalyst for a new approach to understanding the major transformations in the history of life.

In the 1800s zoological expeditions roamed the world exploring continents, mountains, and jungles. They described thousands of new minerals, species, and artifacts. Exploration vessels often had a naturalist on board whose job it was to collect and study the species, rocks, and landscapes that the ship encountered. The eminences of the day were the people who were in a position to analyze and publish on the specimens that arrived on the docks and at the train stations of London, Paris, and Berlin.

If ever a zoologist had a birthright, it was Auguste Duméril (1812–70), a professor at the Museum of Natural History in Paris. Like his father, André, also a longtime professor at the museum, he had a passion for reptiles and insects. Father and son did research together and collaborated to build a menagerie at the museum where they could observe living creatures in addition to preserved ones. Duméril Senior published an influential classification of the animal kingdom, using his son's anatomical descriptions. When André died in 1860, Auguste set out with a vengeance to describe new species.

In January 1864 Duméril received a shipment of six salamanders from a collecting team who were exploring a lake outside Mexico City. The salamanders were large adults, and unlike any adult salamanders known at the time, they had a full set of feathery gills that extended like plumes of feathers from the base of the skull. The creatures even had a keel on their back that extended to a flipper-like tail. The implication was clear: with gills and an aquatic body shape, these adult salamanders lived in water.

Unknown to the explorers, the salamanders had long been part of Aztec culture. The species may have been new to science, but in Mexico they were a favored delicacy, often roasted for feasts and special rituals.

Prompted by Darwin's newly proposed theory of evolution, Duméril thought that these aquatic amphibians might provide clues to how fish evolved to walk on land. He placed his new creatures in the menagerie that he and his father had built. Happily, he had both males and females, and after about a year, Duméril got them to mate and produce fertilized eggs. In 1865 the eggs hatched with perfectly healthy juvenile salamanders. Salamanders are easy to care for and, under the right conditions, do not require much food for long periods of time. All was going well with his charges, so Duméril left them alone.

Later that year he looked inside the enclosure. His first thought must have been that someone had fiddled with the cage, because there were now two kinds of salamanders inside. First, there were the parents, the big aquatic adults with gills. But there was another kind living right beside them. These others were also large but looked completely terrestrial, having no gills, no aquatic tail, nothing to suggest they could inhabit water. Looking closely at their anatomy, and comparing them to species already described in the scientific literature, Duméril realized the new creatures had been given a name by scientists years before. They had the exact traits of the genus *Ambystoma*, a well-known species of salamander that were fully land-living.

These animals were so different from each other that, to use

Duméril's two kinds of salamanders

Linnaeus's scheme, they could be classified into two different genera, not just species. It was as if Duméril had put chimpanzees in an enclosure one year and returned the next to find both gorillas and chimps happily cohabiting the cage.

Had a new form of life appeared out of thin air? Had a major transformation happened in Duméril's enclosure in Paris? What magic were salamanders revealing this time?

Developing Stories

For centuries people have looked at embryos with the intuition that somewhere inside the transformation from egg to adult lay clues to the laws that make species different from one another. Indeed, by the time Duméril was puzzling over his salamanders, the development of an embryo, whether of a fish, a frog, or a chicken, was seen as a lens through which to view the biological diversity of every single animal on Earth.

Ever since Aristotle peered inside their eggs, chicken embryos have been objects of fascination. Chicks come in their own container that can be opened much like a window. You can cut a hole in the shell, slide a light along the side of the egg, and pop it under a microscope to see the embryo inside. The embryo begins as a small clump of white cells sitting directly on top of the yolk. Over time it grows, and recognizable landmarks gradually emerge—head, tail, back, and limbs. The process feels like a well-scripted dance. At the very beginning, the fertilized egg undergoes division—one cell becomes two, two become four, four become eight, and so on. As the cells multiply, the embryo eventually becomes a ball of cells. Over a few days the embryo transforms from a hollow ball to a simple disk of cells surrounded by structures that will protect it, provide it with nutrition, and create the right environment for it to develop. From this simple disk of cells emerges an entire creature. No wonder embryonic development has been a source of speculation and scientific investigation.

Charles Bonnet (1720–93) argued that the embryo was, in essence, a small but fully formed miniature being. Its time in the womb was spent growing organs that already existed. These "homunculi," as they were called, were the basis for his view of evolution. Females carried all future generations inside them. The homunculi they carried were able to survive catastrophes, and over time new forms of life would spring de novo from preceding generations of females. The final stage, sometime in the future, would see angels sprout from homunculi in human wombs.

In the century that followed, diverse kinds of embryos were brought to the lab, and new optical technologies were employed to examine them. While Bonnet's idea perished in the face of scientists seeing real embryos, the quest to explain how creatures as different as elephants, birds, and fish are built remained alive.

In 1816 two medical students were among the first to uncover deep insights about biological diversity inside embryos. Both Karl Ernst von Baer (1792–1876) and Christian Pander (1794–1865) were from noble families in the German-speaking regions of the Baltics. Entering medical school in Würzburg, they took a cue from Aristotle and began to look at chicken embryos. Pander incubated thousands of eggs, opened them at different times of development, and put the embryos under a magnifying glass to see how organs formed. He had a distinct advantage over his friend in these early days: coming from a wealthy family, he could afford to build racks to hold thousands of eggs, hire an assistant to draw the embryos, and commission high-quality engravings for publication. Lacking Pander's wealth, von Baer was relegated to the sidelines.

Technological advances worked in Pander's favor—he was able to obtain top-of-the-line magnifying glasses to zoom in on tissues and cells. With an abundance of embryos of different ages, and new lenses with which to view them, he encountered something that no human had ever seen. Embryos in their earliest stages had no recognizable organs; least of all were they the homunculi that Bonnet envisioned. In early stages, embryos did not look like adults, being simple disks of cells sitting on top of the yolk.

Pander wasn't interested only in the external shape of the embryos—he wanted to see what was going inside. Focusing in, he noticed that an embryo started off as a simple disk the size of a few grains of sand. Getting larger through the course of development, the disk eventually became composed of three layers of

Karl Ernst von Baer

tissue, set like sheets one atop another. The embryo at this stage looked something like a disk-shaped cake with three layers.

With thousands of eggs at his disposal, Pander traced what happened to each of those layers as the chick embryos developed and grew from a simple three-layered disk to an adult chicken with head, wings, and legs. He watched organs emerge gradually.

Working under the magnifying glass, and making detailed drawings of every possible stage of development, Pander saw a simple unifying concept in this complex process. The entire organization of the body broke down to these three layers. The inner layer eventually gave rise to the organs of the guts and the glands associated with them. The middle layer transformed to become bones and muscles. And the outer later became the skin and nervous system. To Pander, and to von Baer, who was a friendly spectator to these discoveries, these three layers were an essential organizing principle of the emerging body of the chicken.

Von Baer had a hunch that there were even more insights to come from these layers. Unfortunately, lacking funds, he was unable to do research of his own until a decade later, when he took a professorship at the University of Königsberg. With the income from his new position, he was now able to explore the vast unknown about embryos of different species. His passion occasionally led him astray. To demonstrate the organ that gave rise to mammalian eggs, he sacrificed his director's pet dog. While von Baer is forever associated with the discovery that mammalian eggs come from the follicles in the ovary, lost to history is how the director felt about his experimental methods.

Von Baer asked: What are the mechanisms at work that make one kind of animal different from another? He collected embryos of as many species as he could find, from fish to lizards to turtles. Extracting the embryos from their eggs or wombs, he would keep them in vials with alcohol as a preservative. Then, like his friend Pander before him, he began to see what was common to all animal development and what made each species unique.

Viewing all the different species under the magnifying glass, he made fundamental observations about animal diversity. Every single species began development with three layers: an inner one, an outer one, and a middle one. And as he traced the layers, he found that their fates were exactly the same. The cells of the deepest layer, at the base of the disk, became the organs of the guts and glands associated with them. The middle layer became the kidneys, reproductive organs, muscles, and bones. The outer layer became organs of the skin and nervous system. Pander's original discovery was not only about chickens—it held for animal life more broadly.

This simple observation revealed a universal connection between every organ in every known animal species. Whether the creature is a deep-sea anglerfish or a soaring albatross, its heart comes from cells of the middle layer, its brain and spinal cord from the outer one, and its intestines, stomach, and digestive organs from the inner one. This rule is so fundamental that if you pick any organ in the body of any animal on Earth, you can know which cell layer built it.

Then von Baer made a mistake. He forgot to add labels to a few of the vials that housed different species. Not knowing which species were in which vials, he had to look closely to try to differentiate them. In describing the unlabeled embryos, von Baer said, "They may be lizards, small birds, or very young mammals. The formation of the heads and trunks in these animals is quite similar. The extremities are not yet present in these embryos. But even if they were in the first stages of development, they would not indicate anything; since the feet of lizards and mammals, the wings and feet of birds, as well as the hands and feet of men develop from the same fundamental form."

With his labeling mishap, von Baer saw an order to animal life that unfolds as development continues. Adult bodies often mask profound similarities in early development. While the adults, or even neonates, can look extremely different, in their earliest stages of development they are very similar.

These embryonic similarities run very deep even in their details. The head of an adult fish has few apparent resemblances to that of an adult turtle, bird, or human. But a short time after conception, all these embryos have four swellings that lie at the base of the head. These so-called gill arches, which have a cleft between them externally, develop in any creature that will have a bony skull. Indeed, their presence forms the baseline for the development of different types of skulls. In fish, the cells inside the swellings become the muscles, nerves, arteries, and bones of the successive gills. The clefts that separate the swellings become the gill slits. Even though people don't have gills, we have the swellings and clefts in our embryonic stages. In us, the cells of the swellings become the bones, muscles, arteries, and nerves of parts of the lower jaw, middle ear, throat, and voice box. The clefts never become full slits but seal over to become parts of our ears and throats. We have them as embryos, not as adults.

Example after example—from kidneys and brains to nerves and backbones—made von Baer's case potent and enduring. Sharks and fish have a connective tissue rod running from head to tail underneath the spinal cord. Filled with a jelly-like substance, it forms a flexible support for the body. A human's backbone is composed of vertebrae, blocks of bone separated from one another by intervertebral disks. No rod runs from our head to our hips. Yet our embryos have a fundamental similarity to those of sharks and fish: they have that rod. During development, it breaks up into small blocks that eventually become the inner part of our intervertebral disks. If you've ever ruptured a disk, a painful trauma, you have injured this ancient remnant of development we share with sharks and fish.

Von Baer's observations about the similarity of early-stage embryos of different species caught Darwin's eye. Von Baer's work was published in 1828, and Darwin was aware of it three years later, when he departed on the HMS *Beagle* for his life-changing trip around the world. When he published *On the Origin of Species* three decades later, he offered embryos as evidence for his theory of evolution. To Darwin, the fact that creatures as different as fish, frogs, and people had a common starting point meant they shared a common history. What could be better evidence for the common ancestry of different species than common embryonic stages in development from which they sprang?

Following von Baer's discoveries with embryos, the German scientist Ernst Haeckel (1834–1919), a generation after von Baer, explored a link between embryonic stages of development and evolutionary history. Haeckel trained to become a physician, but he couldn't tolerate seeing sick patients, so he went to Jena to study with a leading comparative anatomist. His life changed when he read and met Charles Darwin.

Haeckel scoured the animal kingdom for embryos and produced more than one hundred monographs describing and illustrating embryonic stages of diverse species. He envisioned a seamless connection between art and life: the diversity of life was a form of art to him. He produced some of the most beautiful color lithographs ever made. His voluminous renderings of corals, shells, and embryos reflect an age when careful anatomical drawing bridged science and aesthetics. Embryos in particular were celebrated not only for their beauty but for the way they connected to Darwin's new theory. Haeckel, always quotable, coined a phrase linking the two that was to linger like an advertising jingle for many who studied biology in the twentieth century: "Ontogeny [development] recapitulates phylogeny [evolutionary history]."

Haeckel's claim was that animal embryos, as they develop, track the creature's evolutionary history: a mouse embryo looks successively like a worm, a fish, an amphibian, and a reptile. The mechanism that produces these stages lies in the way new features arose in evolution. He proposed that new evolutionary features were added to the end stages in development; for example, amphibians arose by adding amphibian-specific features to the end stages of the development of a fish ancestor, reptile features to those of amphibians, and so on. Over time, according to Haeckel, this process resulted in embryonic development tracking evolutionary history.

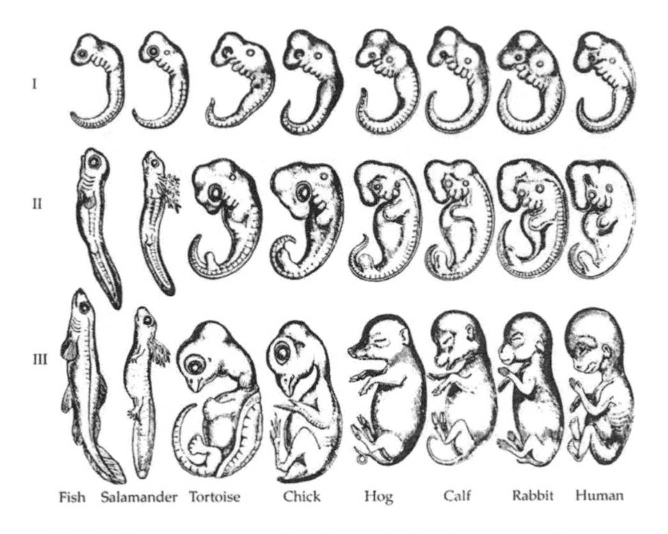

Haeckel's comparison of embryonic development of different species. This was an influential yet controversial figure. Some argued that he overemphasized the similarities among embryos and took liberties with his diagrams.

Who needed intermediate fossils to trace life's history if, as Haeckel supposed, it could be read in embryos? Haeckel's notion was so influential that it launched people on expeditions to obtain embryos of different species. On one of these expeditions, Robert Falcon Scott's 1912 Antarctic expedition to reach the South Pole, three members became consumed with the search for emperor penguin eggs. The explorers thought that the embryos of emperor penguins, which were considered primitive at the time, would hold clues to how birds arose from reptiles. Somewhere in their embryonic development would be a stage that looked like their reptilian ancestor.

In the middle of an austral winter, the three crew members departed on a monthlong sledge trip from their base to Cape Crozier, where the penguins had their rookery. In pitch-darkness,

Apsley Cherry-Garrard (right) after returning from his worst journey to get penguin eggs

with temperatures dropping to minus sixty degrees Fahrenheit, the three nearly died several times when their tents blew away or when they slipped into crevasses. One of them, Apsley Cherry-Garrard, wrote in his classic travelogue, *The Worst Journey in the World*, that the team managed to return to camp with three penguin eggs. The expedition later lost Scott and four crew members, including two of Cherry-Garrard's compadres from the penguin trip, in their tragic and failed attempt to reach the pole. Afterward Cherry-Garrard returned to Britain and tried to deliver the eggs to the British Museum. The museum made him wait in the hall for several hours as they decided whether to accept the eggs. Reluctantly, they took them, but as Cherry-Garrard wrote to the museum head later, "I handed over the Cape Crozier embryos, which nearly cost three men their lives and cost one man his health. . . . Your representative never even said thanks."

The reason the museum was reluctant to accept the eggs was that in the interval between the expedition's departure for the pole and Cherry-Garrard's return, Haeckel's recapitulation theory had been widely discredited and, in addition, the supposed primitive nature of penguins had been challenged by new discoveries. Haeckel had stirred such interest in embryology that he sowed the seeds for his own downfall. Eager to find evolutionary history in embryos, scientists studied embryonic development in diverse species. For the most part, von Baer's idea of a similarity among embryos of different species held up, albeit with some exceptions. But the new data didn't support Haeckel's recapitulation theory; in fact, it did quite the opposite. At no stage of embryonic development could an ancestor be seen. Human embryos may look in some ways like fish embryos, as von Baer suggested, but never in their development do they look like one of our ancestors, whether it is a fish with legs or an Australopithecine; nor does a bird embryo look like *Archaeopteryx* during its development.

Haeckel's idea was wrong, but it guided the research of countless scientists. It lingers even in some quarters today, despite the fact that it has not been a topic of scientific inquiry for over a century. Perhaps Haeckel's most lasting influence was on the person who loathed his idea the most.

The Axolotl

Walter Garstang (1868–1949) so despised Haeckel's idea that he developed a critique that led to a new way of thinking about life's history. He had two lasting, if eccentric, pursuits—tadpoles and verse. When he wasn't doing science on larvae, he was writing limericks and jingles about them. His passions came together in a book published two years after his death, *Larval Forms and Other Verses*, where he transformed a career of scientific research into poetry.

"The Axolotl and the Ammocoete" may not sound like a promising title for verse: it refers to a salamander (axolotl) and a tadpole-like animal (ammocoete). But the idea expressed in the poem changed the field and defined research programs for decades. Garstang's notion explains not only what happened in Duméril's magical enclosure but also some of the revolutions that made our own presence on this planet possible. To Garstang, larval stages weren't simple detours of development; they were rich with artifacts of the history of life and potential for its future.

Most salamanders live in water for much of their development on the undersides of rocks, on fallen branches in streams, or at the bottom of ponds. Their larvae hatch with a wide head, small flipper-shaped limbs, and a broad tail. A cluster of

A portrait of Walter Garstang that appears at the beginning of *Larval Forms and Other Verses*

gills projects from the base of the head like a bunch of feathers extending from the shaft of a feather duster. Each of the gills is broad and flat, maximizing the surface over which it can take up oxygen from the water. With their finlike limbs, broad flipper-like tails, and gills, these creatures are clearly built for life in water. Axolotl larvae are born with very little yolk in the egg, meaning they must feed voraciously if they are to grow and develop. The broad head serves as a huge suction funnel: when they open their mouths and expand their gapes, water and food particles get pulled inside.

Then, at metamorphosis, everything changes. The larvae lose their gills and reconfigure the skull, limbs, and tail, changing from an aquatic creature into a land-living one. New organ systems allow the creatures to inhabit a new environment. Feeding is different on land from in water. The head structures that were so useful in sucking prey into the mouth in water don't work in air. So the creatures reconfigure their skulls to allow their tongues to flop out and pull in their prey. A simple switch affects the entire body—gills, skull, circulatory system. The shift from water to land, something that happened over millions of years in our own fishy past, happens over a few days of metamorphosis in these creatures.

After encountering these striking changes to the salamanders in his menagerie, Duméril traced their entire life cycle. These salamanders—the axolotls of Garstang's verse—normally metamorphose from aquatic larvae into terrestrial adults. But, as Duméril later found, they don't always—they have two different pathways, depending on the environment they experience as larvae. Salamanders that grow in a dry environment will undergo metamorphosis and proceed to lose all their aquatic traits to become terrestrial adults. Those reared in wet environments never undergo metamorphosis and grow to look like big aquatic larvae, with a full set of gills, a flipper-like tail, and a wide skull best suited for feeding in water. Unknown to Duméril at the time, the specimens he obtained from Mexico were big adults that did not undergo metamorphosis because of their wet habitat. Their offspring, which developed in the dry menagerie, underwent metamorphosis and lost all their aquatic larval traits in the process.

The magic that happened in Duméril's enclosure was a simple shift in the ways animals develop. We now know that the major trigger for metamorphosis is a spike in the levels of thyroid hormone in the bloodstream. The hormone triggers some cells to die, others to proliferate, and still others to transform into different types of tissues. If the levels of hormone stay flat, or if the cells cease to respond to it, then metamorphosis will not happen, and the creatures will keep their larval features into adulthood. Changes in development, even small ones, can produce coordinated modifications of the entire body.

Picking up on Duméril's work, Garstang promoted a general principle: small changes in the timing of development can have huge consequences for evolution. Let's say there is an ancestral sequence of developmental stages. If development is slowed or stopped early, then the descendants will look like juveniles of their ancestors. In salamanders, this alteration would cause their bodies to look like aquatic larvae, retaining external gills and having limbs with fewer fingers and toes. Alternatively, if development is extended or sped up, new exaggerated organs and bodies emerge. Snails develop their shells by adding whorls during development. Some snail species have evolved by extending the time of development, or by developing faster. These descendant snails have a larger number of whorls than their ancestors. The same kind of process explains a wide variety of large or

Salamanders can slow or stop their development and change their bodies dramatically.

exaggerated organs, whether the antlers of elk or the elongated necks of giraffes.

Tinkering with embryonic development can make dramatically new kinds of creatures. Ever since Garstang, scientists have generated taxonomies of the ways developmental timing can be altered to produce evolutionary changes. Slowing the rate of development is a different process than terminating it early; each mode can produce similar outcomes—juvenilized descendants—but the causation is different. The same relationship between causation and outcomes holds for the process that can produce exaggerated or larger features when development is sped up or extended.

In searching for different causes, scientists have probed for genes that may control these events or for hormones, such as thyroid hormone, that may trigger them. This approach to development and evolution, known as heterochrony (from the Greek *hetero* meaning "other" and *chronos* meaning "time"), has become its own subfield of research. In more than a century of comparing embryos and adults of diverse species, zoologists and botanists have shown how changes in the timing of developmental events can make new kinds of bodies in animals and plants.

Garstang himself revealed one stunning example from our own history—when our ancestor was a worm.

The Ammocoete

Garstang's poem "The Axolotl and the Ammocoete" explored two of the most classic revolutions that happened by retaining larval features in the course of evolution. The axolotl shows the extent of changes that occur when development is stopped early. The larva, a transitory stage in the life of a salamander, becomes the endpoint of development. The ammocoete is a small wormlike animal with a backbone. While it may live by quietly sucking mud at the bottom of rivers and streams, its biology tells a much larger story.

Over two thousand years ago, Aristotle identified and described hundreds of species of snails, fish, birds, and mammals. He distinguished animals with blood inside (*enhamia*) from those without (*anhamia*). This distinction broadly correlates to what we recognize today as vertebrates and invertebrates. There are two kinds of animals on the planet, those with backbones and those without them. The bodies of people, reptiles, amphibians, and fish are fundamentally different from those of flies and clams. At the core of vertebrate architecture is what von Baer saw in fish, amphibians, reptiles, and birds: every vertebrate at some stage of its embryonic development has gill slits, a cartilage rod that supports the body, and a nerve cord running above it. As we've known since von Baer, some of these traits may be obscured or lost in the adult body, but they are present at an embryonic stage. The speculation has been that the ancestor of vertebrates was a simple wormlike creature that had these three features.

For Garstang and many of his contemporaries, the key question was how this body plan came about. Were there invertebrate animals that had these traits in some form? If so, how did our branch of the tree of life evolve from them? Earthworms don't have gill slits or the cartilage rod in either their embryos or adults. Nor do insects, clams, starfish, or most any other animal without a backbone. The answers came from a most unexpected animal, one that is shaped like a lump of ice cream and spends almost its entire life attached to rocks in the ocean.

There are about three thousand known species of sea squirt in the world's oceans. With some species shaped like a scoop of ice cream topped by a large chimney-shaped structure, they sit, sometimes for decades, attached to the rocks beneath the surface, simply pumping water. Water gets pulled into a big tube at the top and goes through the body, only to be expelled by a tube that projects from the center of body. As water travels through their bodies, they filter particles out to feed. Sea squirts take any number of shapes, from clumps to twisted tubes, but they have no obvious head, tail, back, or front. You could not imagine a creature less likely to tell the story of one of the most basic events in human history.

Garstang was interested in their larvae. He explored something remarkable, first seen by Russian biologists in the late 1800s: when sea squirts hatch from the egg, they are free-swimming tadpoles. Not until metamorphosis do they sink to the bottom of the water column and attach to rocks. If there is any tadpole that could capture the imagination, this is it. It swims about looking nothing like the adult. With a big head, it maneuvers by flexing its long tail back and forth. Inside the body a nerve cord runs along the animal's back, and a connective tissue rod extends from head to tail. It even has gill slits at the base of the head. The three great features that are the basis for the putative ancestor of animals with backbones are present in the larval sea squirt.

Then larval sea squirts lose it all, or at least the features that from our anthropocentric viewpoint are important. After a few weeks, the tadpole swims to the bottom of the water column. As it descends, it loses the tail, the nerve cord, and virtually all of the connective tissue rod; it modifies the gill slits to become part of the pumping apparatus. It attaches to the rocks to spend the rest of its days in one place pumping water. A tadpole, a creature

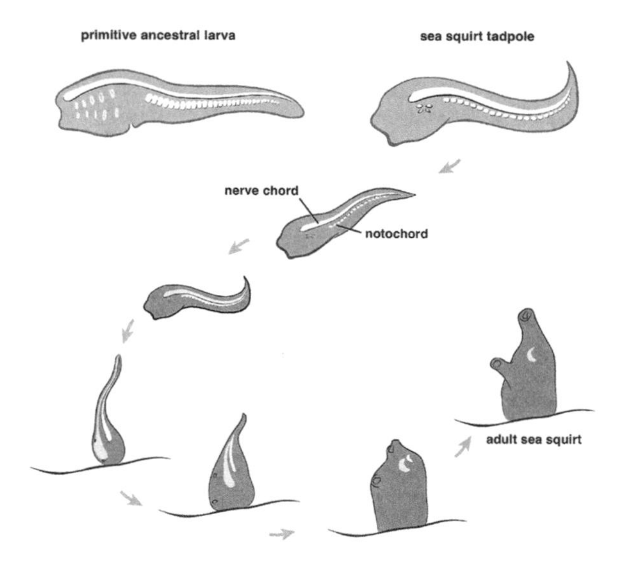

A sea squirt looks like an amorphous lump but begins its development with many traits that we share.

with our vertebrate body plan, transforms into something that has been mistaken for a plant.

Garstang proposed that a shift in the timing of development was a first major step in the transition from invertebrate to vertebrate. An adult human or fish has no resemblance to a sea squirt; many would find the comparison insulting. But its larvae contain the essence. The ancestor of all vertebrates came about by stopping sea squirt development early, freezing the traits of the larval stage, and letting the creature grow to adulthood with them. The result was an adult that looks like a tadpole of its sea squirt ancestors. This creature, with the nerve cord, connective tissue rod, and gill slits, in a freely swimming animal, would become the mother of all fish, amphibians, reptiles, birds, and mammals.

A Picture of Change

Examples of evolution occurring as a result of changes in the timing of developmental sequences abound; it is hard to pick up certain scientific journals nowadays and not see papers on it. Arguably one of the most seminal examples is also one of the most personal.

The years spanning 1820 and 1930 were an age of big ideas in biology. Von Baer, Haeckel, Darwin, Garstang, and countless others looked to anatomy, fossils, and embryos for rules to explain why animals appear the way they do. At the same time, the mechanisms that brought about the diversity of life were becoming known.

In this intellectual milieu, the Swiss anatomist Adolf Naef (1883–1949) rose through the academic ranks, studying with some of the leading lights of the day in Switzerland and in Italy. His goal, as he described it to his brother in 1911, was to formulate "a general science of the form of organisms, a subject on which I have a number of new ideas."

Naef was a meticulous anatomist who knew the impact a good picture or image could have in making a scientific argument. His life, however, was defined in many ways by argument. As he wrote to his brother, "My demeanor alienates most people; some appreciate me all the same, others will have to accept me as pure intellect. I expect enemies rather than friends." In an earlier letter, he asserted that "there exists in Switzerland no abundance of first-rate intellects which is what I take myself to be." With this type of attitude, Naef was never able to find employment in Switzerland, so he spent most of his career at a post in Cairo.

While in Cairo, Naef developed a theory of biological diversity that reflected the philosophy of Plato two thousand years before. In his *Republic*, Plato held that all physical objects were but physical manifestations of ideal essences, the timeless universals that underlay all diversity. The diversity of all objects, from drinking cups to houses, could, to Plato, be boiled down to a metaphysical essence from which each physical manifestation was derived. Naef applied this idea to biological diversity. In his idealistic morphology, as it became known, animals, too, have an essence within their physical diversity. And for Naef, this essence was seen in similarities among animals during embryonic development.

Naef's theoretical framework has largely been forgotten, replaced by new data from genetics and evolutionary relationships. His most enduring contribution is, fittingly, one of the images he used in making arguments for his failed theory. The photo shows a neonate chimpanzee and an adult. Struck by the large cranial vault, erect head, and small face of the young chimp, Naef proclaimed that "of all animal pictures known to me, this is the most manlike." He was trying to show how the

Naef's influential photo comparing a juvenile with an adult chimp. The juvenile, likely a taxidermy specimen, is depicted to emphasize its human proportions and posture.

essence of humanity appears in early development. His theory may have been wrong, but this picture was so influential, it continued to catalyze research decades after its initial publication in 1926.

Adult humans have smaller brow ridges than adult chimpanzees, larger brains relative to body size, more delicate skull bones, smaller jaws, and different skull proportions. But in each of these features, humans are more similar to juvenile chimps than they are to adult chimps. Development also appears to have slowed, as humans have a longer gestational period and childhood than do chimps. By developing more slowly, humans retain many of the proportions and shape of the juveniles of our ancestors, which, as Naef showed, are so very human in many ways.

This notion became a lens through which to view much of human evolution. Paleontologist Stephen Jay Gould and anthropologist Ashley Montagu later observed that essential components of humanity could emerge simply by tinkering with rates of growth and development: couple proportionately large brains for our body size with an extended childhood rich in opportunities to learn, and much of what makes us special may relate to modifying developmental timing. While this explanation of human evolution is simple and elegant, new comparisons reveal that the story is more than an overall slowdown of development. Some human features look like those of juvenile chimps, but others, such as the shape of the legs and pelvis that enable humans to walk on two legs, do not. One hypothesis is that different parts of the body evolve by developing at different rates, the skull evolving by slowing its rates of development while legs and bipedality do the opposite.

Using these and other ideas from anatomy, D'Arcy Wentworth Thompson (1860–1948) postulated a mathematical approach to understanding biological diversity. His goal was to

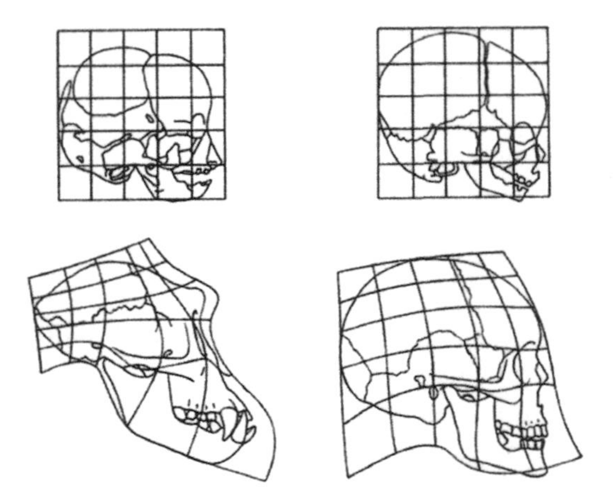

D'Arcy Thompson's grids show how changes in proportion can account for many differences in the shape of skeletons, as in this case of humans and chimps.

reduce the differences in shape among creatures to simple diagrams and equations.

Written during the First World War, his book *On Growth and Form* spawned many an anatomy career, with its diagrams that were as simple as they were influential. Place a Cartesian grid over the skulls of a baby chimpanzee and a baby human, making the lines go through similar points in each. Then do the same for the adult skulls, making the grid lines go through the same locations that they went through in the babies.

The result is that the neat grid lines in the neonates become warped in the adults, and the deformation reflects changes in shape. This depiction reveals that during growth, the chimpanzee and human begin with relatively similar proportions, but then the chimp's cranium shrinks in relative size while the lower face and brow ridges expand. In humans, the cranium expands while the face expands only moderately. In Thompson's view, differences between humans and chimps result less from new organs than from shifts in proportions of different parts of the body, much like those produced by slowing down or speeding up rates of development.

One Cell to Rule Them All

Altering the timing of events is but one way of making evolutionary changes by tweaking embryonic development.

Ever since the days when Pander studied embryos under a magnifying glass, we've known that the development of diverse body parts is often highly coordinated. A simple shift in the working of a single cell, or a handful of them, could cause alterations to many parts of the adult body. The effect can be seen even in the names we give developmental maladies. Hand-foot-genital syndrome, for example, is a genetic mutation that affects the behavior of cells early in development. That single change affects the size and shape of the fingers, the configuration of the feet, and the tubes that carry urine from the kidneys. With such wide-ranging impacts from small alterations, changes in the kinds of cells that build bodies may hold clues to some of the revolutionary changes we see in history.

To understand this way of evolving, we need to return to sea squirts. As Garstang showed, and as recent DNA evidence has confirmed, one crucial step in the transformation from invertebrate to vertebrate occurred when larval features of sea squirts were retained to make a vertebrate ancestor. This tadpole-like adult had the basic architecture upon which the vertebrate body is built. But there was another step in the origin of vertebrates.

Vertebrates such as humans and fish are not simply larval sea squirts. From the bony skeletons that support the body, to the fatty myelin sheaths that surround nerves, to the pigment cells that lie in skin, all the way to the nerves that control the muscles in the head, vertebrates have hundreds of features that invertebrates do not. A list of all the differences between invertebrates and vertebrates would include organs and tissues from head to tail. Clearly more than changes in the timing of developmental stages was involved with this transformation.

Raised by a mother who was widowed soon after her birth, Julia Barlow Platt (1857–1935) was a biology prodigy. After graduating from the University of Vermont in three years, she attended Harvard University, where she dove in to study the embryos of chicks, amphibians, and sharks. True to her talent and ambition, she set an audacious goal for herself. The head is arguably the most complicated part of the body; not including teeth, the human skull has almost thirty bones, and there are more in the skulls of fish and sharks. The head's anatomical complexity derives from the fact that these structures are supplied by a tangle of special nerves, arteries, and veins that are situated in a relatively small container. Platt traced adult structures, such as jaws and cheekbones, to their earliest embryonic stages. Perhaps by studying how skulls develop, she could distill essential similarities hidden in the adult body. Whether she knew it or not, she was entering one of the most contentious areas of science.

The academic climate of the time was not friendly to women pursuing higher degrees. After struggling at Harvard, Platt found a more open culture in Europe and entered a graduate program in Germany. Thus began a nomadic existence that would take her across Europe to the Marine Biological Laboratory in Woods Hole, Massachusetts. There Platt met O. C. Whitman, the director of the marine lab, and she followed him to the University of Chicago, where he was later to become chair of the zoology department.

In Whitman's freewheeling lab, ambitious young scientists were treated as junior colleagues and could follow their own leads for research. In this setting, Platt thrived. Using specimens she collected at Woods Hole and techniques Whitman taught her in Chicago, she looked at head formation in salamanders, sharks, and chicks. Her reason was as much technical as anything else: these creatures have big embryos that develop inside an egg, making them easy to see and manipulate.

With Whitman, she developed a laborious but accurate method to trace cells during development. Her starting point was the three embryonic layers that Pander and von Baer had discovered in the 1820s. By the time of Platt's work these three layers were taken almost as a biological axiom: cells of the inner layer form guts and associated digestive structures, the middle layer the skeleton and muscles, and the external layer the skin and nervous system. Platt noticed that the cells of the outer and middle layers differed in size and in the number of granules of fat inside. Using this distinction as a marker, she traced small groups of cells from each layer to see where they ended up in the skull. This approach allowed her to see which head structures came from which layer.

According to the dogma of the time, all the bones of the salamander skull should have come from the middle layer. But Platt's fat granules showed her something else altogether. Some of the bones of the head, even the dentine of the teeth, were coming from the outer later, which supposedly was restricted to becoming skin and nervous tissue. To some, this finding was heresy. Leading researchers set themselves in opposition to her. One prominent scientist wrote, "An examination of a number of series and stages has not enabled me to find the slightest evidence in favor of Miss Platt's conclusions." This was just one voice in a chorus of criticism, which, for a young female researcher in the 1800s, could end a career before it started.

Fortunately for Platt, Anton Dohrn (1840–1909), the influential leader of the Stazione Zoologica in Naples, picked up on her research idea. He was originally skeptical of her discovery, but her careful analysis persuaded him to use her markers to study development in sharks. He wrote, "I fully agree with the views that we owe to Miss Platt. . . . It goes without saying that I also make this conversion and now oppose all critical papers and remarks directed against Miss Platt's findings."

In Platt's time, there was little room for women on science faculties, particularly individuals who spouted notions that confronted entrenched orthodoxies. Not being able to find employment in science, she moved to Pacific Grove, California, to set up her own small research group. Still making discoveries, she wrote to David Starr Jordan, president of the newly formed

Stanford University. Desperate for a job in science, and knowing she had made fundamental breakthroughs, she ended her letter saying, "Without work, life isn't worth living. If I cannot obtain the work I wish, then I must take up the next best."

Unemployed and feeling unemployable in science, Platt left the field. She brought her strong will and fierce independence to new challenges. Within a short time, she was elected the first female mayor of Pacific Grove, where she led an effort to set up a sanctuary saving Monterey Bay from overdevelopment. Residents and visitors to Monterey today can feel the impact of Julia Barlow Platt.

Platt died in 1935 and did not live to witness her vindication almost forty-three years after her first paper on the subject. Following in her footsteps, researchers developed refined methods

Julia Platt after her term as mayor of Pacific Grove, California

to mark cells during development. They injected dyes into the cells of embryos and traced where they ended up in later stages. In another technique, researchers took patches of cells from a quail and transplanted them into a chicken embryo at different times of development. Since quail cells can be distinguished easily from those of chicks, the scientists could see which organs emerged from them. Both techniques confirmed that the structures in the head that Platt had studied did not come from von Baer's middle layer. The cells start off on the developing spinal cord and migrate to the gills to make gill bones.

The discovery that cells migrate between layers is not just a quirky asterisk to the organization of cells in the three-layer embryo—it has deeper implications for our understanding of how new structures arise. Those cells break off from the developing spinal cord to migrate all over the body of the embryo. Once at their new sites, they make tissues. They become pigment cells, myelin sheaths of the nerves, and bones of the head, among many others—all the features that are unique to vertebrates. The big shift in the transformation of Garstang's ancestral animal to a vertebrate, involving novel tissues across the body, can be traced to the origin of a single type of cell, a new derivative of von Baer and Pander's outer layer. Platt was right in ways she never could have envisioned. The cells she identified were a precursor to all the tissues that make vertebrates special.

Garstang had shown that a first step in the origin of backboned creatures came from a change in the timing of development, retaining larval sea squirt features into adult descendants. Platt's discovery helped reveal the next transition, the origin of a new kind of cell. In both cases, complex changes across different organs and tissues can be distilled to simpler shifts in development. Altered timing at one step and the origin of a new type of cell at another can produce a new body plan.

Of course, these observations raise questions: How do changes in development happen? What kinds of biological shifts can cause embryological development itself to evolve?

Living things do not inherit skulls, backbones, or cell layers from their ancestors—they inherit the processes to build them. Much like a family recipe passed along and modified during each generation, the information that builds bodies has continually changed over millions of years as ancestors pass it on to descendants. Unlike a recipe used in a kitchen, the one that builds bodies anew in each generation is written not in words but in DNA. To understand biological recipes, then, we need to learn to read a whole new language and see new kinds of antecedents in the history of life.

3

Maestro in the Genome

"We have discovered the secret of life!" With that apocryphal boast, Francis Crick (1916–2004) ushered James Watson into the Eagle Pub in Cambridge and the rest of us into the age of DNA. One year later, in 1953, the scientific announcement of the discovery had a very different tone. In the pages of the august journal *Nature*, Watson and Crick open their article with a dry British understatement that others have emulated in the years since. Their discovery, they noted, "has novel features which are of considerable biological interest."

Both announcements heralded something later generations have come to take for granted. The duo modeled the structure of DNA, showing that it exists as double strands that, when separated, can make proteins or copies of themselves. With this trick, the molecule can do two remarkable things—hold the information to make proteins that build bodies and pass that information along to the next generation.

Watson and Crick, following the work of Rosalind Franklin and Maurice Wilkins, found that individual DNA strands are composed of sequences of other molecules, set like beads on a string. Each of these molecules, known as bases, can be one of four types, typically designated A, T, G, and C. One DNA strand can have a series of billions of bases, forming strings like AAT-GCCCTC or any combination of the four letters.

It is a humbling thought: much of who we are resides in the order of molecules in a chemical strand. If you think of DNA as a molecule that contains information, it is as if we have millions of supercomputers in every cell. Human DNA is composed of a chain of roughly 32 billion bases. That strand is broken up into chromosomes, wrapped, and coiled to sit inside the nucleus of each cell. Our DNA is packed so tightly that if unwound, connected, and stretched out, each strand would be about six feet long. Each of our trillions of cells contains a tightly wound six-foot-long molecule coiled to one-tenth the size of the smallest grain of sand. If you uncoiled the DNA from each of the four trillion cells in your own body and put them end to end, your personal DNA strand would run almost to Pluto.

When sperm and egg unite during conception, the fertilized egg ends up with DNA from both parents. Hence genetic information flows from generation to generation. Our own DNA comprises contributions from our biological parents, our parents' DNA from their biological parents, and so on, ever deeper into the past. DNA forms an unbroken connection among living things through time. One of Darwin's great insights can be deployed to translate this simple notion of a family lineage to an even broader history. The molecular implication of his idea is that if we share a common ancestor with other species, then there should be a continuous flow of their DNA to our own. Just as our DNA passes from generation to generation, from parents to offspring, so, too, should it pass from ancestral species to descendant species over the four-billion-year history of life. If true, DNA is a library that resides within each cell of every creature on the planet. Locked in the order of those As, Ts, Gs, and Cs would be a record of billions of years of changes in the living world. The trick has been to learn how to read it.

With influential relatives that included famous anatomists, philosophers, artists, and a surgeon, Émile Zuckerkandl (1922–2013) was born in Vienna into a world of ideas, science, and art. As the Nazis came to power in Germany, his family sought refuge in Paris and Algiers. Family friends connected Zuckerkandl with Albert Einstein, who, using his influence, obtained an entrée for young Émile to study in the United States. The move took Zuckerkandl to the University of Illinois and laboratories there studying the biology of proteins. With an interest in oceans, he gravitated to marine stations in the United States and France during summers. There he became fascinated by crabs and the molecules at work when they grow and molt from tiny embryos to full-grown adults.

Zuckerkandl was entering biochemistry at a propitious time. In the late 1950s, scientists at the National Institutes of Health, as well as Francis Crick himself, were starting to decipher what the strings of As, Ts, Gs, and Cs meant. Each DNA sequence carries the instructions to make yet another sequence of molecules. Depending on the circumstances, a DNA sequence can be used as a template to make a protein or it can make copies of itself. To build a protein, the string of As, Ts, Gs, and Cs gets translated into a sequence of another type of molecule: amino acids. Different strings of amino acids, in turn, make different proteins. There are twenty different kinds of amino acids, and any one of them can reside at any point in the sequence. This code can produce an enormous number of different proteins. Some simple math: if there are twenty different amino acids that can assemble in any combination, and a protein chain is about one hundred amino acids long, the number of different proteins that can be made is a 1 with 130 zeros behind it. The real number is much higher because the length of the protein in our estimate, one hundred, is relatively small. The biggest protein in the human body, known as titin, consists of a string of 34,350 amino acids.

The mental trick is to remember that DNA is made up of a string of bases, symbolized as letters, that codes for strings of amino acids that in turn comprise proteins. Because different proteins are made up of different amino acid sequences, the DNA sequence encodes for the diverse proteins that help make life anew in each generation.

By the late 1950s, researchers were able to map the sequences of amino acids of different proteins to begin to understand how they work in the body. These discoveries heralded an age in which scientists could study protein structure to understand disease. For example, in sickle cell anemia, diseased red blood cells live for only ten to twenty days, whereas healthy ones can live for almost ten times that. Moreover, sickle cells, as the name implies, have a distinctive shape. This difference causes them to be destroyed in the spleen much more easily than normal red blood cells, which have a disk-like shape. As a consequence, sickle cell anemia, in its most extreme cases, can be fatal by the age of three in almost 70 percent of sufferers. And what is the difference between a healthy red blood protein and a sickle cell one? Only a single amino acid in the string: the amino acid glutamate is replaced by one called valine at the sixth position in the sequence. A tiny difference in the amino acid sequence can have massive ramifications on the protein, the cells in which the protein is found, and the lives of the individuals who have those cells.

Inspired by the power of this new biology, Zuckerkandl turned his attention to species in his marine laboratory. He speculated that when crabs molt from small embryos to full-grown adults, certain proteins are at work. He set out to look at the structures of proteins and how they control crab respiration, growth, and the molting of their shells.

Then his life changed by a form of scientific kismet. Linus Pauling (1901–94), then a Nobel laureate in chemistry, was visiting France and stopped by the marine lab to see some friends. Zuckerkandl, with his love of proteins and crabs, sought out Pauling, more like how a fan would approach a rock star than a scientist looking for a new research project. That interaction would transform Zuckerkandl and, ultimately, science itself.

By the mid-1950s Pauling had uncovered the structure of crystals and the fundamental properties of atoms and molecular bonds, and he had even formulated a molecular theory of general anesthesia. He ended up losing the race with Watson and Crick to uncover the structure of DNA. Later, he would spend considerable effort promoting his theory that vitamin C warded off the common cold and other infections.

Pauling grew up in Oregon and attended Oregon State Agricultural College. His fearless approach to science has made him a hero of mine. I am on the selection committee for a foundation in New York that funds artists and scientists at key moments in their careers. The foundation has been awarding fellowships since the 1920s and has retained every application it ever received. Its offices on Park Avenue are a treasure trove of letters, files, and applications of Nobel laureates, novelists, dancers, and academics of all stripes. A colleague there knew of my interest, and when I came to work one morning, I saw an old crinkled file waiting on my desk. It was Pauling's application from the 1920s. At the time applications required college transcripts and doctors' notes, items we would never request today. I took particular interest in his transcript from Oregon State. His record was distinguished by its highs and lows. As expected, he had As across the board in geometry, chemistry, and math. His work in "camp cookery" merited an undistinguished C. Gym was an ongoing string of Fs for years. In his second year, Pauling established one of the top grades in his class in a required course on "explosives." He ultimately won two Nobel Prizes: after receiving the award in chemistry in 1954 for understanding proteins, he won the peace prize in 1962, for his work against nuclear testing. Pauling's As in chemistry and explosives in college augured well for his future life.

After a short conversation, Pauling saw something special in Zuckerkandl and invited him to move to Caltech. But Pauling's offer came with strings attached. Pauling did not have a lab of his own at the time because he was away most days working on his antinuclear activities. Pauling set Zuckerkandl up with a colleague whose lab was equipped to do biochemistry experiments. When Zuckerkandl broached his idea of working on crab proteins, Pauling waved that aspiration aside. For over a decade, Pauling had been interested in how nuclear radiation could affect cells. One target of this work was the protein hemoglobin, which ferries oxygen in the blood from the lungs to the cells of the body. Pauling suggested, to put the term mildly, that young Zuckerkandl give up the aspiration to understand crabs and instead spend his time thinking about hemoglobin. While the shift derailed Zuckerkandl's plans, the advice was prescient.

Zuckerkandl explored the hemoglobin proteins of different species using some of the era's technologies, which were quite limited. He couldn't sequence amino acid composition of the proteins of different species, so he extracted them and used relatively simple methods to assess their overall size and electric charge. With the safe assumption that proteins having generally similar amino acid sequences should have similar weights and electrical charges, he used these easily obtainable measurements as proxies for their overall similarity.

Zuckerkandl found that human and ape hemoglobins were more similar to each other in size and charge than they were to the hemoglobins of frogs and fish. This simple measurement held, for him, the glimmer of something important. He speculated that this similarity between human and ape proteins could be the result of evolution: the reason human and primate blood proteins were similar was because they are closely related. When he showed his initial result to the head of the laboratory, Zuckerkandl got the cold shoulder. The professor was an ardent creationist and would have none of this evolution talk in his laboratory. Zuckerkandl was welcome to work there, but the boss would have nothing to do with any publication that suggested that people and monkeys were related to each other. The door seemed to close for Zuckerkandl just as he saw a glint of success.

Then luck struck. Pauling got an invitation to contribute a paper to a *Festschrift* for another Nobel laureate, his close friend Albert Szent-Györgyi. *Festschriften* are books or special issues of journals produced to honor the retirement of a valued colleague. They typically contain papers celebrating a career in science contributed by friends and longtime colleagues. The key point is that virtually nothing important ever appears in these volumes, because the papers are usually remembrances sprinkled with slivers of new data. These volumes are not often peer reviewed; hence they can hold long pages of adulation for the honoree or data that authors couldn't publish anywhere else. Knowing these facts, and wanting to honor his friend, himself a very bold scientist, Pauling had an idea. He approached Zuckerkandl with the idea of writing "something outrageous."

This offbeat aspiration fueled one of the classic scientific papers of the twentieth century.

The timing was ripe for doing something bold in biochemistry. By the time Zuckerkandl entered Pauling's orbit in the late 1950s, the amino acid sequences of different proteins were becoming available, and Pauling's lab had access to the data. Today's DNA sequencing was still a long way off, but sequencing the amino acid string of different proteins was possible, if tricky and slow. Pauling was acquiring sequences of the proteins of gorillas, chimps, and people, among others. Armed with this new information, Zuckerkandl and Pauling were ready to attack the fundamental question: What do the proteins of diverse animals tell about their relationships? Zuckerkandl's initial results, using crude analyses of size and charge, implied that proteins might tell quite a bit about history.

A century before anybody knew about DNA and the sequences of proteins, Darwin's ideas had made specific inferences about them. Darwin speculated that if creatures shared a genealogical tree, then the amino acid sequences of proteins of humans, other primates, mammals, and frogs should reflect their evolutionary history. Zuckerkandl's initial experiments hinted that this could be the case.

Hemoglobin turned out to be an ideal subject for this research. All animals use oxygen in their metabolism, and hemoglobin is the blood protein that carries oxygen from the respiratory organs, either lungs or gills, to the body's other organs. Zuckerkandl and Pauling compared the amino acid sequence of the hemoglobin molecule in different species and were able to estimate how similar the proteins were.

Each new species Zuckerkandl and Pauling added to their analysis brought Darwin's prediction into ever clearer focus.

The sequences of humans and chimps were more similar to each other than to cows. And all these mammalian hemoglobins were more similar to each other than to those of frogs. Zuckerkandl and Pauling confirmed that they could decipher the relationships among species, and the history of life more generally, from proteins.

The pair took their idea one step further in a bold thought experiment. What if, they speculated, proteins evolved at constant rates over long periods of time? If that were true, then the more proteins of two species differed from each other, the longer the time those species have been evolving independently from a common ancestor. By this logic, the reason proteins of humans and monkeys are more similar to each other than they are to those of frogs is that humans and monkeys share a more recent common ancestor with each other than either does with frogs. This makes sense given what we know from paleontology—the primate common ancestor of humans and monkeys would be more recent than the amphibian one they share with frogs.

If, as Pauling and Zuckerkandl speculated, proteins evolved at a constant rate, you could use differences in the sequence of proteins to calculate the time that these species shared that common ancestor. (See pages 231–32 for a discussion of the method.) Proteins in the bodies of different species could serve as a kind of clock for understanding evolution: no rocks or fossils would be needed to tell time in the history of life. This idea, so utterly outrageous when it was first proposed, is now known as the "molecular clock" and is used in many instances to calculate the antiquity of diverse species.

Zuckerkandl and Pauling were devising an entirely new way to infer the history of life. For more than a century, the history of life was deciphered by comparing ancient fossils. But now, by knowing the structure of the proteins of different animals, Pauling and Zuckerkandl could assess evolutionary relationships. This insight heralded a bonanza: bodies contain tens of thousands of proteins. The proteins of different species could be as informative as fossils. But these fossils aren't in rocks—they lie inside every organ, tissue, and cell of every body of each living animal on the planet. If you knew how to look, you could uncover the history of life in any well-stocked zoo or aquarium. The history of all creatures was now knowable, even those for which the fossil record had yet to be unearthed.

DNA passes from generation to generation containing the information to make proteins and thereby bodies. Individuals and their bodies may come and go, but the molecules form an unbroken connection through the ages. The more we dig into that connection, the more we learn about the relationships between all living things.

With the publication of the *Festschrift* in the early 1960s, Zuckerkandl and Pauling ultimately gave birth to a new field of research using molecules to trace history. But you couldn't have guessed at the future impact of their paper judging from the reaction of the scientific community at the time. "Taxonomists hated it. Biochemists thought it useless," Zuckerkandl recalled on its fiftieth anniversary. Taxonomists, paleontologists—anybody focused on anatomy despised this idea. No longer would these fields have a monopoly on reconstructing evolutionary history. Zuckerkandl and Pauling showed that virtually every molecule in the body of living creatures can tell of past events. If paleontologists thought the paper threatened their survival, biochemists could not have cared less about it. Evolutionary studies were, to them, a kind of genteel backwater. In their view, serious scientists worked on protein structure, disease, and function, not on the relationships between people and frogs.

A Molecular Revolution

Chemical reactions and scientific ideas share a fundamental similarity: both typically need catalysts to happen. One person took Zuckerkandl and Pauling's ideas to spawn a community of scientists who approached the history of life with new eyes.

In the early 1960s Allan Wilson (1934–91), a mathematics prodigy from New Zealand, switched to biology and joined the biochemistry faculty at the University of California at Berkeley. This was a time of unrest on campuses generally, at Berkeley in particular, and Wilson became one of the most politically active professors there. He relished disruption in everything he did, so much so that his students described political protests as a kind of group lab meeting.

A simple premise drove Wilson's career until his untimely death at the age of fifty-six. He believed that if you cannot simplify a complex phenomenon into its constituent parts, then you don't understand it. The mathematician in him led him to seek simple rules behind biological patterns and then develop rigorous means to test them. Wilson had a passion for developing bold and outrageously simple hypotheses to explain complex patterns in the history of life. Then he'd try to falsify his idea with as much research as possible. If the idea withstood his own data barrage, it was ready to reveal to the outside world. This approach made Wilson's lab a raucous epicenter for some of the best and the brightest at Berkeley in the 1970s and '80s. His laboratory became an intellectual hothouse with a freewheeling and intense attitude, attracting talented young students from around the world, many of whom later emerged as luminaries in their own right.

I arrived in Berkeley as a newly minted paleontology Ph.D. in 1987, when Wilson and his team were at the height of their discoveries. My world was centered on rocks and fossils, not on proteins and DNA. Wilson's presentations were already attracting large crowds from across the university, and the battle lines between anatomists and molecular biologists were drawn and deeply entrenched. At one seminar, I was seated with a number of paleontologists who were growing increasingly uncomfortable with each passing slide of Wilson's talk. The crescendo hit when Wilson presented a simple equation, with three variables, that he claimed revealed how fast evolution happens in different species. Seeing this slide, a colleague elbowed me and asked sarcastically, "So most of paleontology fits into that equation?"

For Wilson, the field of evolutionary biology was ripe for his kind of disruption. Zuckerkandl and Pauling's idea of proteins as historical signposts fit his research style perfectly—it was simple and could be put to the test with new data. Animals have many proteins, proteins were becoming known with great regularity, and if there was a strong historical signal in the data, Wilson would not only find it but squeeze every possible inference out of it.

Wilson set his sights high. His question was: How closely are humans related to other primates? If any question was likely to stir up the dust, this was it. And since fossil evidence was relatively sparse for this part of the evolutionary tree, the molecular approach would be particularly meaningful.

Wilson had an almost magical ability to attract students into his orbit, nurture their talents, and help them make transformative discoveries of their own. After attending college in the Midwest, Mary-Claire King went west to study statistics. Arriving in California in the mid-1960s, she lost her drive for math and was hunting for a new intellectual focus. A course on genetics by one of the senior scientists at Berkeley kindled her passion for the field. Sticking her toe into the genetics world, she worked for a year in a lab only to discover that she simply didn't have the touch for lab work. With a scientific career not looking very promising, she took a year off to work with Ralph Nader on consumer activism. Nader invited her to work with him in D.C., a move that would have precipitated a departure from graduate school. She considered the offer as she went to protests at Berkeley. The protests held sway over her time and opened her world to new people and personalities. One of those personalities was Allan Wilson.

After one protest, Wilson convinced King to return to graduate school, if only to earn the Ph.D. as a sheepskin helpful to her work in policy. Almost immediately, she was swept into Wilson's data-centered activism in science. But the Wilson lab also presented new challenges for her to overcome: no longer in the realm of equations and numbers, she would now have to learn to work with blood, proteins, and cells.

What made matters even more fraught was that Wilson wanted her to do some sophisticated lab work. Since Zuckerkandl and Pauling had produced their initial work on proteins, a number of laboratories were devoting themselves to understanding which living apes are our closest relatives and how long ago our species diverged from them. Wilson and his group believed answers would come from getting as much new data as possible. In classic Wilsonian fashion, King decided to look not just at hemoglobin but at every protein she could get her hands on. A concurrence of signals in many different proteins should constitute a robust evolutionary signal. King and Wilson received chimpanzee blood from various zoos and human blood from hospitals. If King didn't have a knack for laboratory work, she was going to have to find one: chimpanzee blood clots extremely fast, so she would have to work quickly or develop new methods. In the end, she did both.

King decided to use a rapid method to test the differences between proteins. The idea is a simple version of the one that Zuckerkandl had used a decade before. If two proteins differed in their sequence of amino acids, then their weights would differ also. Moreover, being composed of different amino acids means that they would carry different electrical charges. From a technical standpoint, if you put those proteins in a gel suspension to hold them and then ran a current through the gel, the proteins would migrate across to one edge, attracted by the charge. Similar proteins would migrate at the same speed, but proteins that were different would not. You can envision the gel as a kind of racetrack, where the charge would set the race in motion. Similar proteins would go a similar distance in a similar time. The more different they were, the farther apart their runs on the gel would be.

King launched her work still unsure of her skills. And now, to make matters worse, Wilson went off to Africa, leaving her largely on her own during his yearlong sabbatical. She would try to telephone him every week to review her data, but she was largely unmentored for days at a time.

From the start, things did not go well. King managed to extract the chimp and human proteins and put them on the gels. She ran the gels, but the chimp and human proteins moved almost the exact same distance for almost every protein. She wasn't seeing any meaningful differences between humans and chimps. Had she extracted the proteins correctly? Was she running the gels poorly? Her hopes for a breakthrough seemed doomed.

During their regular conferences, King would share her data with Wilson, who would, in typical fashion, hammer her results with questions on technique as if he were still in Berkeley. No matter how hard he hit her work with every conceivable criticism, the result stood. The protein sequences of humans and chimpanzees were nearly identical. And it wasn't just one protein that was telling the story, it was more than forty of them. In fact, King wasn't flailing around aimlessly; she was revealing something fundamental about genes, proteins, and human evolution.

King then compared humans and chimps to other mammals. And here the importance of her discovery came into clear focus. Humans and chimpanzees are more similar genetically than two different species of mouse are to each other. Nearly identical species of fruit fly differ from one another genetically more than humans and chimps do. Humans and chimpanzees are, at the level of proteins and genes, almost identical.

King's gels revealed a deep paradox. The anatomical differences between humans and chimps, including the essence of our human uniqueness—bigger brains, bipedalism, proportions of the face, skull, and limbs—weren't deriving from differences in the proteins or genes that code for them. If the proteins and DNA that make those molecules are largely the same, then what was driving the differences? King and Wilson had a hunch but not the technology to test it.

Recent science has confirmed what King and Wilson first saw. Comparing whole genomes, chimpanzees and humans are anywhere from 95 to 98 percent similar.

The next advances didn't come from the hands of a student and her adviser working alone. They would require big science—the kind of science where the results are announced by presidents and prime ministers.

Geneless Genomes

When President Bill Clinton and Prime Minister Tony Blair held a press conference with the heads of rival teams sequencing the human genome—the publicly supported one led by Francis Collins and the private one directed by Craig Venter—they had only a very rough draft of the genome to announce. Despite the hoopla, at the time of the announcement in 2000, large chunks of the genome were missing, and little was known about which parts were important for human health and development.

The initial outcomes of the Human Genome Project had less to do with genomes than with technology. The race to sequence the human genome set off a technological frenzy that continues to the present day. Gordon Moore famously predicted in 1965 that microprocessing speed would double every two years. We feel the results of that increase with every purchase of digital devices: computers and phones have gotten ever more powerful and cheaper with each passing year. Genome technology has smashed even those rates of progress. The Human Genome Project took more than a decade, cost over $3.8 billion, and involved rooms full of machines. Today, there is an app for sequencing, and handheld gene sequencers are already on the market.

Once the human genome was mapped, those of other species emerged annually. Genomes are now announced so rapidly that the pace is limited only by the frequency at which scientific journals get published. We've had the mouse genome project, the lily genome project, the frog genome project—projects on everything from viruses to primates. At first it was a big deal to have a genome project published; the results would appear in A-list journals to great fanfare in the press.

Nowadays, unless there is some important biological process or health issue at stake, new genomes get published with barely a mention.

While the luster of genome papers has faded, they continue to be a bonanza that would have delighted and enthralled Émile Zuckerkandl, Linus Pauling, and Allan Wilson. Armed with the genomes of flies, mice, and people, we can now look to them to ask central questions about life: How are species related, and what makes each one different?

Each of us is made up of trillions of cells—muscle, nerve, skeleton, and hundreds of others—working together, all packed and connected in just the right way. The flatworm, *Caenorhabditis elegans*, gets by with only 956 cells. If that is not surprising enough, consider this: despite the vast differences in number of cells and complexity of organs and body parts, humans and worms both have the same number of genes, roughly twenty thousand. And worms are just the beginning. Flies, too, have about the same number as we do. In fact, animals are true pikers compared to plants such as rice, soy, corn, and cassava, all of which have almost twice as many genes. Whatever is driving the evolution of complex new organs, tissues, and behaviors in the animal world isn't coming from having more genes.

Even weirder is the organization of the genome itself. Remember our mantra: genes are strings of bases that are translated into a sequence of amino acids, and those amino sequences code for proteins. In essence, genes contain the molecular template for proteins. When a gene sequence is published, authors are required to make the data publicly available and deposit the information in a national computer database. After decades of work on genes, these repositories are burgeoning with sequences from thousands of genes from thousands of species. You can now sit at your desktop, type in a sequence, and see which gene from what species matches it. When you compare a whole genome to the genes in these databases, you can get a picture of what genes are inside by looking at the matches. In genome after genome published over the past two decades, one observation is completely inescapable: genes are rare in genomes. If genes are the part of the genome that codes for protein, then most of the genome doesn't seem to be involved in making them. Gene sequences that code for proteins compose less than 2 percent of the human genome. That leaves some 98 percent with no genes at all in it.

Genes are but islands in a sea of DNA. With rare exceptions, this pattern holds for species from worms to mice. If most of the genome does not contain genes that code for proteins, then what does it do?

Bacteria to the Rescue

After serving in the French resistance during World War II, two French biologists, François Jacob (1920–2013) and Jacques Monod (1910–76), started work on bacteria to understand how they digest sugar. If any question seemed more esoteric and less related to the human condition, this was it.

Jacob and Monod showed that the common bacterium *Escherichia coli* can digest two sugars in its environment, glucose and lactose. The bacterial genome is relatively simple. Long stretches hold genes that contain the information to make the proteins that digest each sugar. When glucose is abundant, and lactose is rare, the genome makes the protein that digests glucose. When the reverse is true, the genome makes the one that digests lactose. While this state of affairs may seem simple and obvious, it was the basis for a revolution in biology.

The scientists discovered two components in the bacterial genome. In the first, the genes contain the information about the structure of each protein that digests the two different sugars. These are the As, Ts, Gs, and Cs that get translated into the sequences of amino acid strings that comprise a protein. Flanking the genes are other shorter strings of As, Ts, Gs, and Cs that don't code for protein at all. When another molecule attaches to this stretch, it turns the gene on or off. This is the second component. Think of these shorter strings as molecular switches that control when a gene will be active and make a protein. In bacteria, genes and the switches that control their activity lie next to each other in the genome. Depending on which sugar is present, a molecular reaction controls which gene is active and, in turn, which protein is made.

Jacob and Monod discovered that the bacterial genome is a biological manufacturing process that makes proteins in the right place and time. There are two components: genes that code for proteins and switches that tell the genes when and where to be active. For this work, the pair won the Nobel Prize for Physiology or Medicine in 1965.

In the decades since Jacob and Monod's Nobel, the twofold organization of the protein manufacturing process has been revealed to be a general feature all genomes. Animals, plants, and fungi all have genes that code for proteins and molecular switches that turn the genes on and off.

Their discovery provides clues to understanding what makes cells, tissues, and organs distinct. A human body is essentially a highly organized package of four trillion cells of two hundred different kinds, organized as tissues, from bone and brain to liver and skeleton. Cartilage tissue is composed of cells that make collagen, proteoglycans, and other constituents that combine with water and minerals in the body to give cartilage its pliant yet supportive properties. The constellation of proteins that make a nerve cell are different from those in cartilage, muscle, or bone.

Here's the rub: every single body cell contains the same sequence of DNA, derived from the fertilized egg that started it all. The DNA inside a nerve cell is virtually identical to that in cartilage, muscle, or bone. If each cell has the same genes inside, then the differences among different cells come from which genes are active making proteins. The kinds of switches that Jacob and Monod discovered become essential to understanding how the genome builds different cells, tissues, and bodies.

If the genome is thought of as a recipe, then genes code for the ingredients, and the switches contain the instructions about when and where to add each ingredient. If 2 percent of the genome is made up of genes that make proteins, then part of

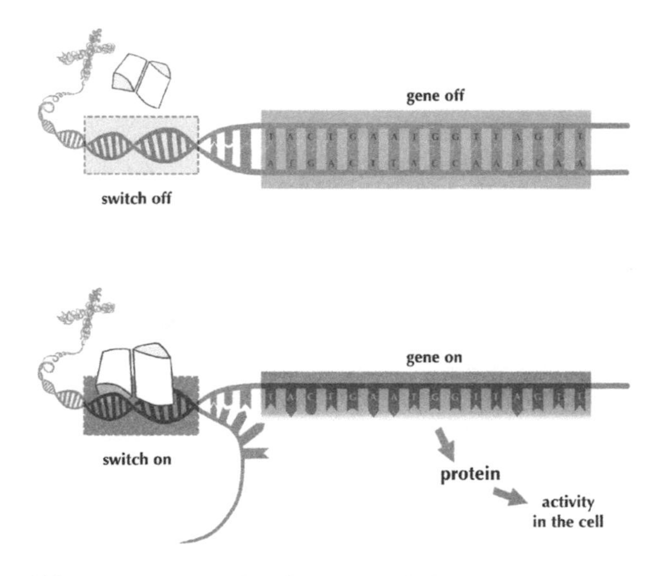

When a genetic switch is flipped, usually by proteins attaching to it, a gene becomes active and makes a protein.

that other 98 percent contains the information that tells genes when and where to be active.

But how does the genome build a body? How does it produce changes to species in the history of life? Nobody knew it at the time of the Human Genome Project, but the small number of genes and their rarity in the genome were only the tip of the iceberg of surprises to come.

Fingers Point the Way

Sailors once believed that six-toed cats could bring good luck on ships. These so-called mitten cats were thought to make better mousers because their broad paws could balance them while at

Hemingway cats, or mitten cats, have broad paws with six or more digits.

sea. Stanley Dexter, a sea captain, had a litter of these cats and gave one to his pal Ernest Hemingway, who was living in Key West at the time. This kitten, Snow White, gave rise to a lineage of six-toed cats that thrives to this day at the Hemingway estate. Besides being a highlight for tourists, these cats have played a role in a new conception of the workings of the genome.

People, too, occasionally have extra fingers and toes. About one in every thousand people is born with an extra digit in the hand or foot. In an extreme case, in 2010, a boy in India was born with thirty-four digits. Extra fingers can appear on the thumb side or the pinky side, or in split and forked fingers. Additional digits on the thumb side, known as preaxial polydactyly, are particularly important biologically.

In the 1960s scientists working on chicken eggs were probing how wings and legs are made in the embryo during development. Limbs emerge from the embryo's body as tiny buds, looking like small tubes. Over a few days—the number varies by species—the bud grows, bones begin to form, and the growing end becomes shaped like a broad paddle. Digits, wrists, and ankle bones form inside this expanded surface.

Scientists discovered that by removing or moving the cells inside the paddle area, they could tweak the number of digits that form. If they excised a small strip of tissue from the terminal end, development of the limb stopped. If they cut out this strip during early development, the embryo formed a limb with few digits or none at all. If they extracted the strip at slightly later stages, the embryo might lack only a single digit. The stage of development at which the experiment is done matters: early removal has more dramatic effects on the embryo than later removal.

John Saunders and Mary Gassling from the University of Wisconsin, for reasons lost to time, extracted a tiny slice of tissue from the base of the growing paddle of a limb bud. This patch is nondescript—nothing about it looks unusual. It sits on the side of the paddle where the pinky will ultimately form. The researchers took this sliver of tissue, less than a millimeter long, and grafted it onto the opposite side of the limb bud, at the base of the paddle where the first digit would form. After sealing up the embryo in the egg, they let it complete development.

The embryo that emerged was a complete surprise. It looked like any normal chick, with a beak, feathers, and wings. But its wings, unlike normal wings with a pattern of three elongated fingers, had as many as six fingers. Something inside that little patch of cells contained instructions to make fingers.

Other labs soon got into the act. In the 1970s a group from England put tiny strips of tinfoil between the patch of tissue and the rest of the limb bud. The wings that emerged had fewer digits than normal. The foil served as a barrier between the patch and other cells. The implication is that some compound emanates from that patch of cells, diffuses across the developing limb, and stimulates digits to form. When a foil barrier stops that diffusion, fewer digits develop, and when the barrier is placed at a different point in the limb, more digits form. But what is the compound that is released?

In the early 1990s three laboratories, working independently, used new techniques to isolate the protein and the gene that makes it. The gene makes a protein during limb development that diffuses across the paddle of the limb bud. As it does so, the researchers found, it tells groups of cells which digits to form. High levels of the protein make a pinky, or fifth digit. Low levels make a first digit, or thumb. Intermediate levels form the digits in between. One of these groups of researchers named the gene *Sonic hedgehog*, a nod both to a gene known as *hedgehog* at work in other species and to a video game popular at the time.

But what tells the gene to make fewer or more digits? Are there switches at work for the *Sonic hedgehog* gene that influence the evolution of digits? Answering this question would be a key for understanding how genes build bodies and how they evolve.

As with most important moments in life and science, this story begins with an accident.

In the late 1990s a team of geneticists in London were inserting snippets of DNA into the genomes of mice to study brain formation. These fragments are part of a little molecular machine researchers make to attach to DNA and to serve as a marker for its activity. Every now and then something goes wrong with this kind of experiment. The fragment can land anywhere in the genome. If it lands in a biologically important part of the genome, a mutant can form. That's what happened with this team's experiment: some of the injected mice developed normal brains but had deformed fingers and toes. In fact, one of the mice had extra digits and very broad paws not unlike Hemingway's mitten cats. The team was able to generate an entire family line of these mutants and, by scientific convention, give them a name. They called them Sasquatch, after the big-footed creature of the paranormal world.

Since their mutants were now useless for the study of brains, the team wondered if any limb biologists might be interested in them. They set up a poster at a scientific meeting announcing their results. Posters at conferences are sometimes thought to contain the B-list of scientific results, as the best ones get presented as talks. But posters also have a social element; people mill around and science gets discussed. It's been my experience that more collaborations begin over posters than after talks.

The poster showed a type of polydactyly that was known to arise from a mutation in *Sonic hedgehog:* the extra fingers were on the pinky side. These mutations happen because *Sonic hedgehog* is turned on in the wrong side of the limb. So the obvious next step was to look at the activity of *Sonic* in the mutants, experiments that the team did to present in their poster. After they accidentally made the mutant, they looked at the tiny developing limbs under the microscope. The activity of *Sonic* in the mutants was abnormally expanded, just as you would have expected in this kind of polydactyly. These observations led to the hypothesis that the mutant Sasquatch had been produced by the snippet inserting into, or very near, the *Sonic hedgehog* gene.

The team didn't attract a limb biologist to their poster, but Robert Hill, a distinguished geneticist at Edinburgh, randomly walked by and saw the photos of the Sasquatch mutant. From that, a new research program began.

Hill's lab had gained renown for understanding the workings of the genome in eye development. Through that work, his team, including the young scientist Laura Lettice, had developed a toolkit to probe the genome to find fragments of DNA. Since they knew the DNA sequence of the snippet, they had to chug through the whole genome looking for where it ended up residing. Lettice was just starting her career and still quite green, but she had the patience and the skill set necessary to pull it off.

The team used a simple trick to identify the general location of the mutation on the strand of DNA. They attached a dye to a small molecule that was complementary to the piece of DNA that made the mutant. The idea was that this sequence would home in on the mutation, attach to it, and voilà, the dye would light up at that location. Since the mutation was affecting the activity of *Sonic hedgehog*, it was likely to be found in one of two places: in the gene itself or in the control region immediately adjacent to it, like the control regions Jacob and Monod had discovered in bacteria.

The reaction did not affect the gene of *Sonic hedgehog*. That area was not lit up by the dye. Whatever was affecting *Sonic hedgehog* in the limb, and causing polydactyly, wasn't a mutation of the gene or, correspondingly, a change in its protein. The team concluded, as Jacob and Monod had, that one of the adjacent control regions was affected. But when they looked, they saw that this area was completely normal. So if neither the gene nor the adjacent switch was affected, what was the cause of the mutation?

As anybody who has ever tried to recover a model rocket on a windy day knows, you can waste a lot of time looking for something nearby when you should be looking really far afield. Hill, Lettice, and the team started trudging through the entire genome until they saw the signal. The snippet inserted was almost a million bases away from the *Sonic hedgehog* gene. That's an enormous amount of genetic real estate between the site of the mutation and the site of the *Sonic* gene. Thinking they must

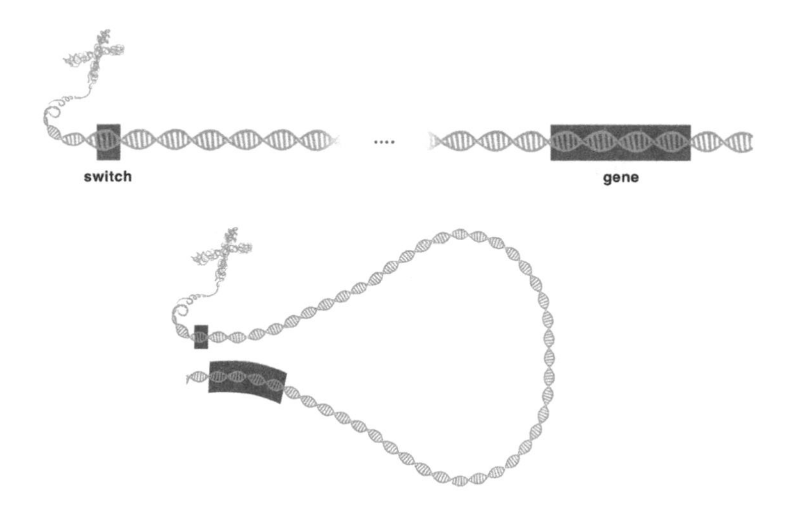

Some genetic switches are located far from the gene they control. DNA is always looping, folding, and contorting to open and close, bringing switches back to the neighborhood of their gene to turn them on and make a protein.

be wrong, they repeated the process and reanalyzed the results. But try as they might, the result stood. A small region one million bases from the gene somehow controlled the activity of *Sonic hedgehog*. It was like finding the switch for a light in a living room in Philadelphia on a wall in a garage in suburban Boston.

Maybe changes to this remote site were the source of the extra digits? The team tracked down every six-fingered person or cat they could find—polydactylous patients in Holland, a child in Japan, even Hemingway's cats—and examined their DNA. And in every single one, they found a slight mutation in that region one million bases away from the *Sonic hedgehog* gene. Somehow, a little mutation at the far end of the genome causes a change in the activity of *Sonic*, turning it on broadly across the limb, leading to additional fingers and toes.

While sequencing the pattern of As, Ts, Cs, and Gs in this special region, they found this stretch of DNA to be very distinctive. It is about fifteen hundred bases long, and its sequence is comparable among different creatures. People have the region in the exact same place as mice do, about one million bases away from the gene. So do frogs, lizards, and birds. It is present in everything with appendages, even in fish. Salmon have it, as do sharks. Every creature that has the *Sonic hedgehog* gene active in the development of its appendages, whether limbs or fins, has this control region almost one million bases away. Nature was telling scientists something important with this odd genomic arrangement.

Changing Recipes

At first glance, it is a wonder that polydactylous cats and people even survive to birth. *Sonic hedgehog* does not merely control limbs during embryonic development; it is a master gene controlling the development of the heart, spinal cord, brain, and genitals as well. *Sonic* is like a general tool that development pulls from its toolkit to make diverse organs and tissues. Accordingly, a mutation in the *Sonic hedgehog* gene should affect every structure where it is active; mutants would have deformed spinal cords, hearts, limbs, faces, and genitalia, among other organs. But what kind of animal would likely arise from a mutation in the *Sonic hedgehog* gene? Since so many aberrant tissues would likely be produced by a mutation in *Sonic hedgehog*, the answer would certainly be a dead one.

But the way *Sonic hedgehog* is controlled during development ensures this outcome doesn't happen. Why? Mutations in the limb-control region only affect limbs. That's why polydactylous people with this kind of *Sonic hedgehog* mutation have normal hearts, spinal cords, and other structures: the switch that controls the activity of the gene is specific only to a particular tissue, leaving the rest unaffected.

Imagine a house with many rooms, each with its own thermostat. A change to the furnace will affect the temperature in every single room, but changing a single thermostat will affect only the room it controls. The same relationship is true for genes and their control regions. Just as a change in the furnace will affect the entire house, an alteration in a gene, and the protein that is produced, can affect the entire body. A global change would be catastrophic, producing dead ends in evolution. But since the genetic control regions are specific to tissues, like a thermostat in a room, a change in one organ won't affect any others. Mutants can be viable, and evolution can work.

Two kinds of genomic changes can play a role in evolutionary transformations. In the first, changes in genes can cause new proteins to form. A mutation in the sequence of As, Ts, Gs, and

Cs in DNA could bring about a change in the amino acid chain that makes protein. If the DNA mutation causes a different amino acid to form along that string, then a new protein can be produced. This clearly happens in many of the major proteins of the body, such as the hemoglobin genes that Zuckerkandl and Pauling studied. The key point is that a change in a protein can affect the body everywhere that protein is found.

The second type of genomic change can occur in the switches that control the activity of genes. After seeing Bob Hill's work, a lab in Berkeley wanted to find out whether the *Sonic hedgehog* switch was involved in limb evolution. They started with snakes, since they lack limbs altogether. When the region of the genome that holds the switch was removed from a snake and placed inside a mouse, the mouse's limbs failed to form digits. Over time it appears that snakes acquired mutations in the switch that controls their ability to form limbs. The *Sonic hedgehog* protein in snakes is completely normal, as are their hearts, spinal cords, and brains. The change to the switch active in limbs meant that only the activity of *Sonic* in limbs changed.

This genetic trick holds the clues to general mechanisms for revolution in evolution. If the past decade and a half of research is any indicator, changes in the switches that control gene activity are behind major shifts in evolution of the bodies of vertebrates and invertebrates for organs as different as skulls, limbs, fins, fly wings, and worm bodies, among many others. In case after case, evolutionary transformations are less about changes in the genes themselves than in when and where they are active in development.

David Kingsley, a geneticist at Stanford, spent nearly two decades studying the tiny threespine stickleback, a fish that lives in oceans and streams around the world. Sticklebacks come in a variety of shapes: some have four fins, others two, and still others show different body shapes and color patterns. This diversity makes the stickleback a powerful system in which to explore how genetic changes can make fish different from one another. Using genomic technology, Kingsley has been able to show the exact regions of DNA that underlie most of these changes. Virtually every one is a switch that controls gene activity. The fish with only two fins has a gene with dramatically altered activity that inhibits the activity of a gene necessary for the development of the hind fin. He showed that the change was not to the gene but to the switch that controls the activity of the gene. Guess what happened when he took the switch from a fish that has four fins and put it into the ones that normally have only two? He resuscitated hind fins by making a four-finned mutant from two-finned parents.

We now have the technology to scan the entire genome to see where genes and their control regions reside. Control regions lie everywhere in the genome; some are close to the gene, while others, such as those for *Sonic hedgehog*, are far away. Some genes may have many control regions influencing their activity, others only one. However many there are, and wherever they may lie in the genome, there is an elegance, indeed a mystery, to how this molecular machine works.

New microscopes that allow us to see DNA molecules themselves also let us see what happens as genes turn on and off.

For a gene to become active, a molecular game of Twister needs to happen. Inactive regions of the genome are tightly coiled upon themselves, bundled around other small molecules to fit inside the nucleus. These regions are closed off and so are relatively inert. Before a region of the genome can become active, it needs to uncoil and open itself up to make a protein.

These are only the first steps in a finely choreographed dance that turns genes on and off. For a gene to activate, its switch needs to contact other molecules and attach to an area adjacent to the gene itself. These attachments trigger the gene to make a protein. In the case of *Sonic hedgehog*, the switch needs to fold a very long distance to initiate the activity of the gene. So here are the full steps of the dance that goes on when genes turn on: the genome opens, revealing the gene and its control region, parts attach, and a protein is made. This happens in every cell, with every protein.

A six-foot-long string of DNA is coiled until it is smaller than the size of the head of a pin. Conjure the image of it opening and closing in microseconds, writhing and turning to activate thousands of genes every second. From the moment of conception and throughout our adult lives, our genes are continually being switched on and off. We begin as a single cell. Over time, cells multiply, while batteries of genes are activated to control their behavior to form the tissues and organs of our bodies. As I write this book, and as you read it, genes are switching on in all four trillion of our cells. DNA contains many supercomputers' worth of computing power. With these instructions, a relatively small parts list of twenty thousand genes can build and maintain the complex bodies of worms, flies, and people using control regions spread across the genome. Changes to this incredibly complex and dynamic machine underlie the evolution of every creature on Earth. Always coiling, uncoiling, and folding, our DNA is like an acrobatic maestro, a conductor of development and evolution.

This new science speaks to Mary-Claire King's struggles to find differences between human and chimp proteins four decades ago. She and Allan Wilson foresaw the importance of genetic switches in the title of their 1975 paper, "Evolution at Two Levels in Humans and Chimpanzees." One level was at the genes, the other at the mechanisms that control when and where genes are active. Major differences between humans and chimpanzees lie not in the structure of their genes and proteins but in the switches that control how they do their jobs during development. Seen in this way, the gulf between creatures that look as different as humans and chimpanzees, or worms and fish, becomes smaller at the genetic level. If a protein controls the timing or pattern of a developmental process, then changes to when and where that protein is active can have big effects on the bodies of adults.

Changes to the switches that control gene activity can affect embryos and evolution in a myriad of ways. If, for example, proteins that control brain development are turned on for a longer duration or in different places, the result can be larger and more complex brains. Tweaking the activity of genes can bring about new types of cells, tissues, and, as we'll see, bodies.

4

Beautiful Monsters

MONSTERS LOOM LARGE in speculations about the workings of nature. In the centuries before Darwin, the word *monster* had an almost technical meaning. Natural philosophers and anatomists crafted taxonomies to describe two-headed goats, multilegged frogs, and conjoined twins. In the sixteenth century, many thought these deformities came about as the result of too much seed during conception or from a pregnant woman's wandering thoughts.

A new science was heralded in the 1700s when the German anatomist Samuel Thomas von Sömmerring (1755–1830) surmised that monsters reflect alterations in normal development rather than mystical causes. They were, in his words, "disruptions of the generative force." On the title page of his monograph on the subject in 1791, he depicted duplicated human heads: stillborn infants with two complete heads sprouting from the neck, and others with duplications of only the face. In his view, each case represented an alteration of normal development at different stages. Complete duplicate heads came about from disruptions of early stages of development, while incompletely fused faces arose from later ones.

A few decades later, Geoffroy Saint-Hilaire proposed that *monstres*, a term he used frequently, reflected the hidden potential for creatures to transform into one another. After his expedition with Napoleon in Egypt and his encounter with lunged fish (see Chapter 1), Saint-Hilaire spent his days trying to mutate chicken eggs, adding various chemicals to perturb their development. He believed that if he added just the right concoction of chemicals to developing embryos, then he could change one creature into another. Following an early notion that chickens went through a fish stage in their normal development, Saint-Hilaire worked for decades trying to make chicken eggs produce fish hatchlings. That attempt failed, but his son Isidore picked up the mantle and produced a three-volume treatise on congenital anomalies that is still in use today. Isidore developed a taxonomy of birth anomalies, categorizing them by type, organ affected, and degree of anatomical effects. For example, he studied conjoined twins, classifying them according to how many organs were involved and the extent to which their anatomical systems were intermingled. This work formed the basis for later researchers to assess the biological mechanisms, as opposed to supernatural causes, involved in producing anomalies.

With the publication of *On the Origin of Species*, Darwin transformed the study of developmental anomalies. To Darwin, if the motor for evolution is natural selection, then variation among individuals is its fuel. If individuals in a species vary in having traits that look and function differently, and some of those traits enhance the success of those individuals in a particular environment, then over time those creatures and traits should increase. If a trait is harmful, then it will diminish over time. The essence of evolution is variation among individuals. If all individuals in a population are exactly alike, evolution by natural selection could never happen. The differences among individuals are evolution's raw material for natural selection; the more variation, the faster evolution could work. Only with a rich supply of variation, including the type revealed by monsters, could natural selection lead to major changes over time.

One of the champions of the study of variation after Darwin was William Bateson (1861–1926). Like Darwin, Bateson grew up with a passion for natural history. When asked as a youth what he wanted to be, he famously replied that he wanted to be a naturalist, but if he wasn't good enough, he would have to be a doctor. Bateson entered Cambridge University in 1878 as a lackluster student. But Darwin's *On the Origin of Species* had a profound effect on young Bateson. He became energized to understand how natural selection works. For him, answers lay in understanding how species vary: What were the mechanisms that made organisms look different from one another? Reading the work of Gregor Mendel, who discovered the principles of heredity in pea plants, Bateson had an epiphany: variation that was transmitted from one generation to the next was the essence of evolution. He translated Mendel's work into English and invented a new term to describe it: *genetics*, derived from the Greek work *genesis*, meaning "origin."

Bateson, like Geoffroy Saint-Hilaire before him, wanted to classify the ways species and individuals differ. But Bateson had one advantage. Armed with new ideas from the growing field of genetics, he looked for the ways that variation among individuals could influence how evolution worked.

Bateson devoted almost a decade to this study, producing the monumental *Materials for the Study of Variation* in 1894. The book contains a road map of the ways creatures differ from one another and a search for general rules that underlie the production of variation and, ultimately, the path of evolution. In assessing as many species as he could, he described two different modes of variation. One type is a difference in the size or degree of organs, which form a continuous series from smaller to larger. Populations of mice, for example, have differences in the lengths of their appendages, tails, or other organs. This kind of variation can be easily quantified by measurements of length, width, or volume. The other kind of variation is more dramatic, involving the presence or absence of structures. The polydactyly of Hemingway's cats is one example. Normal individuals have five toes, while polydactylous ones have six or more. These cats differ from normal ones in the number of toes they have, not in, say, the length of their bones. This type of variation is of kind, not of degree or size.

The search for creatures with extra organs became a passion for Bateson. He was struck by oddities in nature—extra organs or organs in the wrong place, such as bees with legs where antennae should be, or humans with extra ribs, or males with extra nipples. In these cases, it was as if organs were being cut and pasted throughout the body. A well-formed organ could be duplicated in toto or moved about to different places in the body. These monsters had a mystery to them, and understanding them might reveal general rules about how bodies are built and evolve.

Natural philosophers from the sixteenth century onward had been correct in their view that monsters reflect something essential about the living world. What was needed was the right kind of monster and the scientific tools to understand it.

The Fly

One of the greatest decisions in the history of biology came about when Thomas Hunt Morgan (1866–1945) decided to work on flies. Morgan began his career by researching sea acorns, worms, and frogs, convinced that inside their cells and embryos lay clues to our own biology. Nor did he choose them esoterically or haphazardly; he focused on small aquatic creatures that could rebuild complete body parts after losing them. Planarian worms, for example, are champions of regeneration: cut them in half, let them regrow, and the end result would be two complete individuals. Many creatures—worms, fish, and amphibians—can rebuild after trauma. We can only be jealous of our animal cousins; somewhere along our evolutionary line, mammals lost this ability.

Morgan entered science at a time when much of what we take for granted today was completely unknown. The Czech monk Gregor Mendel discovered that traits can be passed on from generation to generation, but the source of that heredity was a mystery. People had observed cells, but the notion that chromosomes play a role in that process was not known, let alone the existence of DNA.

Implicit in Morgan's science was a fundamental shift in thinking about life, something that undergirds virtually all biomedical research today: diverse creatures, from worms to sea stars, can offer insights into general mechanisms of human biology. His work was governed by the tacit recognition that all creatures on the planet share deep connections.

After a few years of performing experiments on regeneration and describing them in his influential book *Regeneration*, published in 1901, Morgan realized that the tools simply didn't exist for him to make significant progress. He began a hunt for a new research program. At the heart of it all, from regeneration to anatomy, lies heredity—the passing of information from one generation to the next. Learning what drives heredity would be a key to unlocking many of biology's mysteries. Morgan was convinced that insights into genetics would come from finding a creature that bred and grew quickly, was small, and could be maintained in huge numbers in a lab. He ideally wanted a species whose chromosomes, by then proposed but not proven to contain genetic material, could be seen microscopically. This was a pretty long checklist, one that excluded the creature he most wanted to understand—humans.

Unknown to Morgan at the time, an insect taxonomist was on a similar mission, albeit from the opposite side of the problem. Charles W. Woodworth (1865–1940), at the University of California at Berkeley, made it his lifework to uncover the arcane details of insect anatomy, with an eye to classifying flies and other insects. This quest made him an expert in fly biology, so much so that he saw one species, the fruit fly, *Drosophila melanogaster*, as a potential experimental model. Sometime in the early 1900s (the exact year is not known), he reached out to William E. Castle (1867–1962), a biologist at Harvard, and suggested he try some experiments on fruit flies.

Like Bateson, Castle was interested in uncovering the mechanisms of heredity and variation. At the time, Castle was working on guinea pigs to understand how their fur color and body patterns were passed from generation to generation. But guinea pigs were a source of frustration because the females give birth to eight offspring at most and take almost two months to gestate. To study heredity, Castle had to wait months for them to breed enough to make multiple generations. Woodworth's suggestion to work on flies had an obvious attraction; the average fruit fly lives for forty to fifty days, during which time a female can produce thousands of embryos. Castle realized that he could do more experiments on heredity in one month with flies than he could in years with guinea pigs.

Castle switched to working on flies and established methods

to breed and rear them. He published a paper on fly experiments in 1903 that is less memorable for its scientific results than for its impact on the community. Other scientists, including Morgan, saw the beauty and power of studying flies.

Drosophila seems like an unlikely candidate for groundbreaking discoveries. About three millimeters in length, it lives on rotting fruit. Most of us encounter them around garbage as tiny nonbiting flies that annoy by hovering about. But what makes them a pest makes them promising for science.

Morgan's work followed the tradition of monsters, which meant finding and analyzing mutants. Mutants are keys to the functioning of normal genes. A mutant with no eyes reflects a defect in one or more genes that control eye formation. In this way, mutants are lodestars that can be used to identify the genes involved in the development of different organs. Since mutants are rare, Morgan needed to breed thousands of flies to pick up a single mutant. He and his team kept hundreds of breeding colonies of flies and put each individual under the microscope to look for any anomalies.

Unknown to most of us, the fly body that emerges under a microscope is beautifully complex. Seen at medium power, an entire world of bristles, spines, and appendages emerges from their body segments. Morgan's team became familiar with this complexity so that any change, no matter how small, served as fodder for their analysis of new mutants. They spent long hours bent over microscopes looking for flies with any odd trait, perhaps differently shaped wings, novel stripe patterns, or an altered appendage.

Genes, as we now know, are sequences of DNA that are bundled tightly to form chromosomes. Chromosomes sit within the nucleus of a cell, and under the right conditions, they are visible

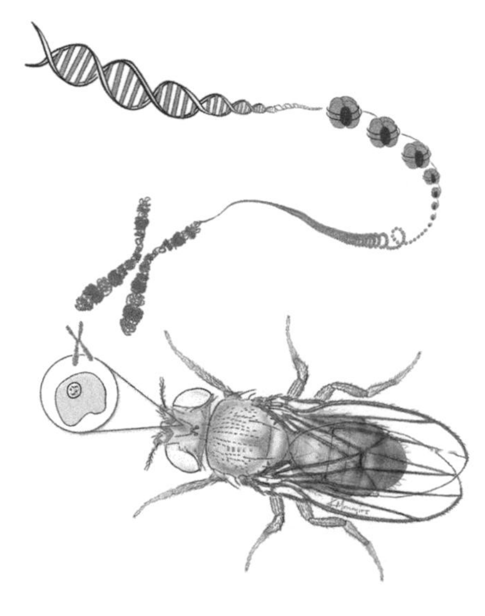

Genes are segments of DNA that are wound and packed tightly into chromosomes that lie within the nucleus of a cell. Notice the banding of the chromosomes.

under a microscope. Morgan knew nothing about DNA, but he could see chromosomes. They became his window into genes.

Morgan devised ingenious ways to try to link the anatomy of mutants to their genetic material. His team found that flies have enormous chromosomes inside their salivary glands. Removing them, and treating them with a red dye obtained from a wild lichen, revealed a series of white and black stripes on the chromosome, some thick and others thin. Morgan then mapped the patterns of white and black bands in both normal flies and those with mutations. By comparing the differences in stripes, he could see the location on the chromosome where the two differed, in essence revealing where the genetic change that made the mutation resided.

The flies fed on rotten bananas, so the Morgan lab was permeated by the smell of garbage. Working there meant spending hours peering into a microscope. Because of these conditions,

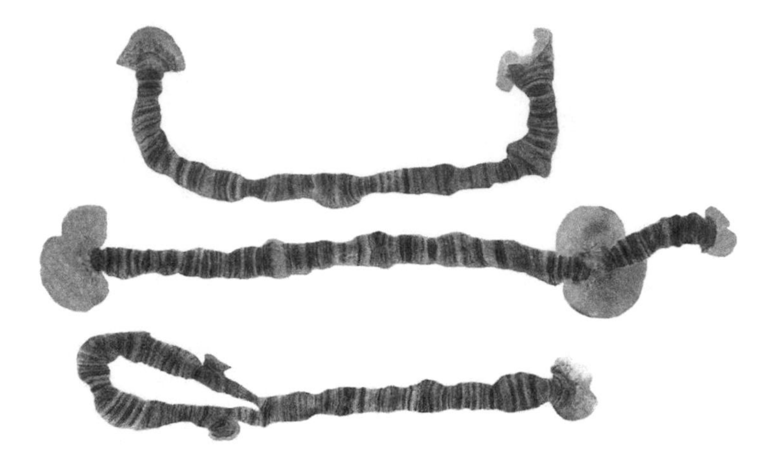

Chromosomes of the midge *Chieronomus prope pulcher,* with black and white stripes

success in Morgan's group called for a special sort of person—one who could, despite all else, stay focused on fly bodies, chromosome bands, and mutants. At stake was one of the biggest questions of life: How is information passed on from one generation to the next?

Morgan's lab was initially in a cramped space at Columbia University, where stocks of flies were stored, bred, and analyzed under the microscope. Known as the Fly Room, the lab would host a who's who of early twentieth-century biologists, as Morgan attracted some of the best and brightest to his lab. After spending fourteen years at Columbia, he moved the entire operation to Caltech in 1928, winning the Nobel Prize in 1933.

One of Morgan's early students possessed a legendary ability to work with flies. Calvin Bridges (1889–1938) had not only the best eyes to discern mutant flies but the patience to sit for hours to find them. Bridges discerned tiny differences among flies that were invisible to others. He brought technical advances too:

switching to a binocular microscope expanded the range of his vision and led to the discovery that flies feed well on agar. The latter was an important change for the lab—no longer would the Fly Room smell like rotten bananas.

With a shock of erect hair that seemingly defied the laws of physics, Bridges was a restless soul. When he wasn't working long hours in the lab, he would often disappear for extended stretches of time. He once emerged with pictures of a new automobile he had designed. Rumors of his amorous trysts abounded, and Morgan disapproved of his private life. The buzz about his affairs meant that Bridges never got promoted to a faculty position at Caltech. When he died in his forties, the word in the lab was that he was killed by the spouse of a jealous lover. Sadly, the truth was just as tragic. Recently a genetics colleague of mine asked his brother, a Los Angeles DA, to dig out Bridges's death certificate. Bridges died from complications of syphilis.

Calvin Bridges and his hair

To the external world, the lab maintained a complete silence about Bridges's personal behavior. But he had had such an impact on Morgan's work that Morgan shared his Nobel Prize winnings with Bridges's family after his untimely death.

While Bridges was known for spotting mutant flies that had subtle differences in coloration, wing shape, or bristle pattern, one of his most famous discoveries was relatively easy to spot. Its difference would have been hard to miss by even a rank amateur. The name, *Bithorax*, says it all—instead of two thoracic segments and wings, it had four. A whole region of its body was duplicated, wings and all.

Bridges drew the fly's body and described its anatomy. Then he did what geneticists do when they find a mutant: he raised the stock and kept it going in the Caltech fly lab. He made a colony of these mutants that could be maintained indefinitely.

Bridges wanted to find the location in the chromosome where the change might have occurred. Using Morgan's technique of staining the salivary chromosomes, he was able to locate a region in the double-winged mutant where the banding was different

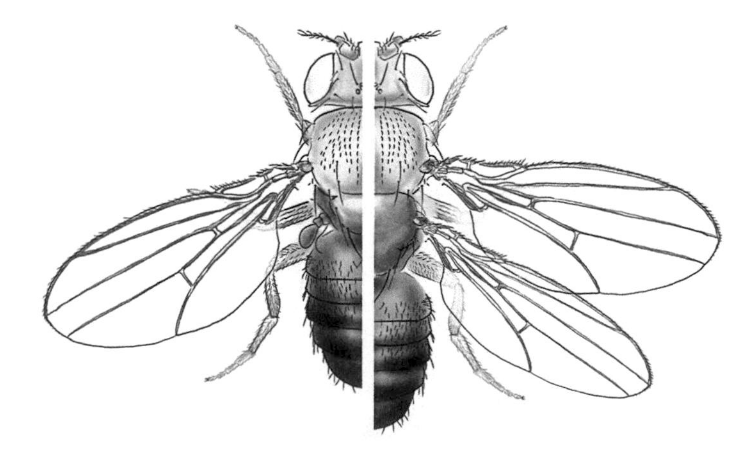

Normal fruit fly on the left, the *Bithorax* mutant on the right

from that of normal flies. The *Bithorax* mutant had happened because of a change in a broad region of the fly's chromosome.

Morgan and Bridges's quest to understand a single trait in flies opened up a new world of challenges and opportunities. They and others showed that various traits in flies are heritable. Some kind of biological material is passed from generation to generation that tells the developing embryo of a fly to place wings in the correct part of the body. Bridges's mutant revealed that this material resided along a stretch of the fly's chromosomes. But what was this material that builds organs and bodies, and how does it do its magic? Could it tell us how bodies are built and how they evolved over millions of years?

Beads on a String

Edward Lewis's (1918–2004) passion for flies was kindled when he saw an advertisement in a magazine. Born in Wilkes-Barre, Pennsylvania, he had an intense curiosity that led him to spend long hours in the local library. Seeing an advertisement for fruit flies, he brought it to the attention of his high school's biology club. The club set up a fly colony, and Lewis began tinkering with flies.

Lewis entered Caltech in 1939, a year after Bridges's death, to learn the tools of genetics that had been pioneered in the Fly Room. He was a quiet man with a very rigid diurnal rhythm: he spent early mornings in the lab, exercised at eight a.m., did more solitary work, had an afternoon lunch at Caltech's famed faculty club, the Athenaeum, then returned to work and played his beloved flute until dinner. He had, like Bridges, a prodigious capacity for sitting for long hours over a microscope working on flies. His favorite time, by all accounts, was the quiet of the lab

Ed Lewis with his flute in the living room of a friend

after dinner. Lewis's work finding and breeding mutant flies was a form of meditation.

The stockroom where Bridges made his great technical advances was still functional and housed the famed *Bithorax* mutants. By the time Lewis started his studies, he knew of the *Bithorax* mutant and also had a hunch about its structure. Since Bridges's map showed that the *Bithorax* mutant spanned several bands on the chromosome, Lewis thought it might be part of a region containing not one but many genes involved in development.

Seeking to isolate the genetic material that made the extra wing, Lewis devised a novel, but time-consuming method to probe *Bithorax*. He spent decades on this work, not publishing a single scientific paper for over ten years as he dedicated himself to *Bithorax*. The six-page article that appeared in 1978 was as revolutionary as it was impenetrable. To understand it all, the paper must be read multiple times, because it is crammed with the insights derived from years of a quiet life with flies.

Lewis had developed a powerful new technique: he would remove a large area of the fly's chromosome and let the fly develop to see the effect on the body in flies lacking this large region. Then he would add small fragments back sequentially, to see those effects on the body. This approach enabled him to determine what individual pieces of a chromosome can do in isolation.

This approach reminds me of a diet that comes in and out of popularity, called a cleanse. People would fast for several days, then add different food groups to their diet sequentially and in combination. By refraining from eating entirely, then adding only dairy products for a few days, they could see how eggs, milk, and cheese affected their energy levels and mood, for example. Then, by fasting and adding foods in different combinations, they could see the interactions, say, between dark leafy vegetables and dairy. Lewis was doing the same with the large region of the chromosome that held the *Bithorax* mutant—he took it out completely, let the animals develop to record the effect, and then added bits back in isolation and in different combinations in other embryos, noting their impact on the flies' bodies as they developed into adults.

Lewis's genetic cut-and-paste revealed that *Bithorax* was not caused by a single gene but a group of many of them. The genes lay in a row on the chromosome, like pearls on a necklace. These genes, he surmised, worked together to build the embryo, and each gene had its own function. But that wasn't the most remarkable thing.

A fly's body is composed of segments from front to back—head, thorax, and abdomen. Each segment carries an appendage:

antennae and mouthparts on the head, wings on the thorax, and legs and spines on the abdomen. Lewis found that each gene in the *Bithorax* region controlled a different segment of the fly's body. One gene placed the antennae on the head, another the wings on the thorax, and another the legs on the abdomen. These genes played a role in building basic body architecture. The front-to-back organization of the body was encoded genetically. And, to everyone's great surprise, the structure of the body was mirrored by the position of the genes on the chromosome: the genes that were active on the head lay on one end, those for the abdomen on the other, and the ones for the thorax in the middle. The organization of the body was reflected in the activity and structure of the genes.

While Lewis's finding was exhilarating, a lot of biology suggested it might pertain only to flies. For one thing, fly segments are different from parts of other animals, such as fish, mice, and humans. Flies lack a backbone, a spinal cord, and other structures seen in bodies like ours. Fish, mice, and people lack antennae, wings, and bristles.

An even greater difference lies in how the fly develops. During development, most animals have millions of different cells, each with its own nucleus. A fly embryo looks like a single cell with many nuclei, like a giant bag of genetic material. You could not imagine a stranger animal than a fly to try to use to say anything about how animals in general develop and evolve.

The Monster Mash

In 1978, when Lewis's paper on *Bithorax* was published, the field of biology was undergoing a technological revolution. In Morgan's day, genes had been a kind of black box—he and his team were able to piece together their effect on the body and their place on the chromosome, but virtually nothing was known about how they worked, let alone that they were regions of DNA.

By the 1980s, a few years after Lewis published his paper, biologists were able to sequence genes as well as see where they were actively making proteins in the body. Mike Levine and Bill McGinnis, working in the lab of the late Walter Gehring (1939–2014) in Switzerland, had access to a mutant fly in which a leg sprouted from the head, where an antenna normally would be. The head developed normally except for the presence of the leg. Much like Bridges's mutant fly with the extra wings, or Bateson's cut-and-paste variations, this mutant shuffled body parts, and it had a defect specific to the head segment.

Using DNA technology that Bridges could not have imagined, Levine and McGinnis were able to isolate the gene responsible for the mutation. Then they made a special piece of DNA to test where the gene was active in development. Recall that

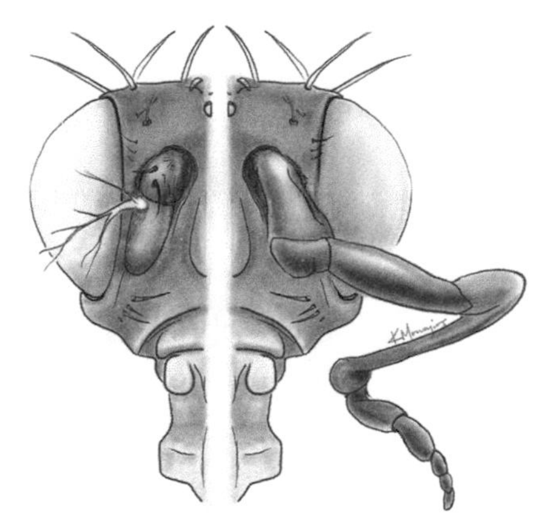

Normal fly on the left, mutant on the right. It was named *Antennapedia* because it sprouted a leg where an antenna should be.

when genes are active, they make proteins. To manufacture proteins, they use another molecule, RNA, as an intermediary. To test where genes are turned on, you need to see where RNA is being made. So the two attached a dye to a molecule that would find the RNA wherever it was in the fly body. When this concoction was injected into a developing fly embryo, the dye would be brought to the places where the gene was turned on, and the stain would be visible in the embryo under the microscope.

The gene of the mutant *Antennapedia*, with the leg growing out of its head, was normally active in a very specific place: the head. Moreover, the gene controlled the kind of organ that formed in the head, whether an antenna or, as in the mutant, a leg. If this situation sounds familiar, it is because it is what Ed Lewis saw in his chromosomal work on *Bithorax* years before. Recall that he saw a series of genes, one after another on the chromosome, each specific to one body segment, each controlling which organ developed there. Maybe this head gene was a harbinger of discoveries to come, one of a group of genes that controlled what was happening in each of the fly's body segments.

The result sent Levine to Lewis's 1978 paper. He began a long interaction with it, reading and rereading it over fifty times, but still, as he said, he "did not completely understand it."

Lewis's paper led Levine and McGinnis to chase one of his major predictions: that there should be a string of similar genes lying next to one another on the chromosome. With the gene isolated, they began a hunt to see if there were any others like it nearby. The technique was crude: they mushed fly bodies into a paste, isolated their DNA, put the mixture in a gel, and added their gene with a dye. The idea was that the gene would act like molecular flypaper and attach to every gene with a similar sequence. The dye would allow them to find and isolate these genes.

The result was unmistakable—there were many other genes like it in the genome. Sequencing each one, Levine and McGinnis found that the dyed genes all had a small stretch of DNA inside that was virtually identical. In a stunning coincidence, Matt Scott, at the University of Indiana, made the same discovery independently.

Now, knowing the sequence of the genes, scientists could apply the same techniques on a larger scale to see where they were active in the fly body during development and where they resided on the chromosome. Using the tricks they had deployed on the mutant that started it all, researchers from around the world found something unexpectedly beautiful: these genes lie next to one another on the chromosome, and each one is active in a different body segment of the fly.

In the midst of this frenzy of experiments, Levine was chatting with a scientist in another lab who pointed out that flies aren't the only animals that have body segments. Earthworms are basically tubes with blocklike segments that run the length of the body. Why not look at them too? Perhaps their genes marked their segments as well.

This casual comment sent Levine and McGinnis running to the garden behind their building to collect every creepy crawly creature they could find: worms, insects, and flies. After extracting each creature's DNA, they probed whether they, too, had genes in a similar sequence. They did. And they didn't stop there. Subsequent research would reveal that the DNA of frogs, mice, and even people had this sequence too.

Subsequent work on worms, flies, fish, and mice revealed universal truths about animal bodies. Versions of the body-building genes of flies turned up virtually everywhere, from worms to people. All these genes were set like beads on a string next to one another on the chromosome. And each gene seemed to be active

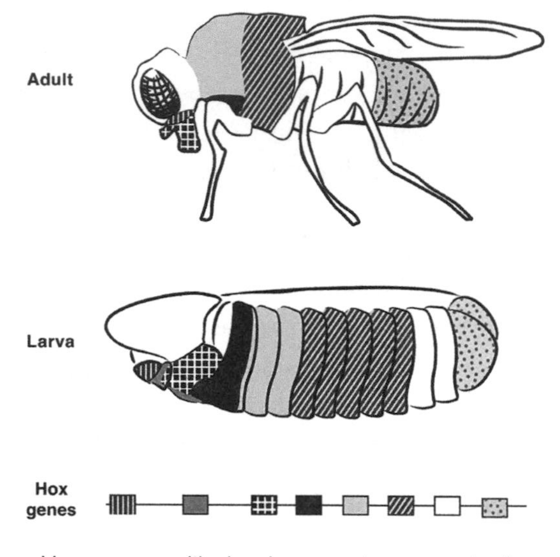

Hox genes, set like beads on a string, are active in the body segments of flies and mice.

in a specific segment of the body—head, thorax, and abdomen. In addition, as Lewis first saw, the position of each gene on the chromosome matched the order of the segments from front to back.

The papers that described these genes were in the stack that kindled my own work in genetics and molecular biology almost four decades ago.

In 1995 the Nobel Prize committee recognized Edward Lewis for opening up a new world of biology. As he accepted the prize, he was classically circumspect. In his acceptance speech, he said prizes were nothing compared to his first loves, "flies and doing science."

The world of bugs, flies, and worms is a mash of creatures with different numbers of segments and different types of appendages emerging from them. Think of a lobster with antennae in front, followed by big claws, small claws, and legs. Each of these appendages emerges from a single segment of its body. In centipedes, each body segment has an identical leg emerging from it. Flying insects have wings instead of legs in certain segments. People have vertebrae, ribs, and limbs that lie along the body. With these genes, scientists could now ask how the basic body architecture of animals developed and evolved.

Calvin Bridges identified the general chromosomal region that made an extra set of wings; Ed Lewis revealed that the region contained many genes, each active in a specific part of the body; and Levine, McGinnis, and Scott showed that those genes are deeply ancient among all animals. A new generation was now inspired and poised to understand how these genes worked.

Cut and Paste

When my children were toddlers at the beach on Cape Cod, they used to find little shrimp-shaped animals in the sand. Poking them, and watching their response, led to their nickname, "jumpies." These creatures, more commonly known as scuds or sand fleas, are about half an inch long, have clear bodies, and usually burrow in beach sand. When provoked, they can contract their bodies and jump a foot or so into the air. The familiar beach variety is only one of the eight thousand known species. All of these species have a remarkable ability to move about by using a diverse array of swimming, digging, and hopping behaviors. They accomplish this with a virtual Swiss army knife of legs: some are large, others small, some face forward, still others face backward. Their name, amphipod, is a Greek reference to having backward- and forward-facing legs: *amphi* means "dual" and *pod*, "leg."

Starting his own independent lab in 1995 in Chicago, biologist Nipam Patel wanted to find a perfect animal to explore how genes work to build bodies. Since amphipods sport so many different kinds of legs, he had a hunch that they could make an excellent creature to study Lewis's genes. He spent years scouring nineteenth-century German monographs to identify the perfect amphipod to bring into the lab. The 1800s were the apex of anatomical illustration and description, and entire rooms in library stacks are dedicated to different groups. Armed with insights from the descriptions and lithographic plates, Patel developed a plan that also fit nicely into his long-standing hobby.

A visit to Patel's house in Chicago meant navigating a giant saltwater aquarium in the center of his living room. Since he was a dedicated amateur aquarist, his experience with the filtration system in his home tank gave him an idea. Keeping the system clean was a regular problem, especially keeping the filter clear of the small invertebrates that collected and grew on it. He couldn't help but notice that amid the grime were small invertebrates burrowing in the muck. Apparently they loved the nutritious particles that flowed by and made it a happy home.

That gave Patel an idea. If tiny creatures liked his small filtration system, imagine the diversity of creatures he might find in the filtered mud of the massive saltwater tanks at Chicago's Shedd Aquarium. These tanks housed sharks, skates, over fifty species of large fish, and even a human docent in scuba gear from time to time. Patel dispatched a graduate student with a bucket to see what he could find in the filtration system. He had a hunch that the muck would harbor robust little animals that he could use in the lab.

The filters at the Shedd proved to be an Eden for small invertebrates. Patel's student spent his days scraping the filters, looking at the creatures that lived there under the microscope.

One of them—an amphipod known as *Parhyale*—was extremely promising for research. It was small, bred fast, and grew to adulthood quickly. It also had appendages, lots of different kinds of them. It looked like a perfect experimental animal. Patel worked to breed them in the lab and get the experiments going. Morgan had used flies to understand the mechanisms of heredity; Patel was determined to use amphipods to explore how genes build bodies.

Soon after obtaining *Parhyale* from Chicago's Shedd, Patel moved to the University of California at Berkeley to establish a research program centered on the creatures. Berkeley, Patel, and *Parhyale* proved to be an auspicious fit, because at Berkeley was Jennifer Doudna, one of the scientists who discovered a new way to edit the genome, CRISPR-Cas. With this technique, scientists can target regions of the genome with two kinds of tools: a molecular scalpel to cut DNA and a guide to bring the scalpel to the right place. In 2013 Doudna and her colleagues from around the world had shown that the DNA of different species could be cut and edited with great precision. Their CRISPR scalpel could be used to cut genes out of the genome. Rearing the embryos would let scientists see the effects of removing one of their genes. Other more complicated experiments involved substituting or editing the sequence of genes.

The power of this technology spawned an idea for Patel: What if you could edit *Parhyale*'s genes to make the genetic activity of one body segment look like that of another? Could you move limbs and body parts around?

Parhyale has limbs along its body length, and each segment of the body contains a different appendage. The front segments of the head have antennae and are followed by segments that contain pieces of the jaw. (We call the jaws and mandibles of invertebrates limbs because, like appendages, they extend from a body segment.) The thorax holds larger limbs, some facing forward with others facing backward. Tiny limbs extend from the abdomen, too, with bushy ones in the front abdominal segments and short stubby ones in the rear.

Six of Lewis's genes are active during the development of the body axis of *Parhyale*. As in flies, different body segments can be identified by the kinds of limbs that develop in them and by determining which of the genes are active in the segment during development. What if you could change the pattern of gene activity in the segments—say, make the thorax segment have the abdomen's genes active inside it? Would that change the kinds of limbs that emerged from the segment? Patel turned genes off one by one, using the gene editing technique developed by his Berkeley colleague.

The elegance of Patel's experiments emerges in the details. Three of Lewis's genes, called *Ubx*, *abd-A*, and *Abd-B*, are active in the rear end of *Parhyale* during development. Their activity in the body marks four regions: one toward the head, where only *Ubx* is active, followed by another where both *Ubx* and *abd-A* are active, one with *abd-A* and *Abd-B* active, and one where only *Abd-B* is active. You could think of each of these four regions as having a genetic address defined by which genes are active inside them. It turns out that the pattern of gene activity corresponds to the kind of appendage that forms. Where only the *Ubx* gene is active, a backward-facing limb forms, the combined *Ubx/abd-A* segment produces a forward-facing limb, the *abd-A/Abd-B* one a bushy limb, and the *Abd-B* segments, a stubby one.

Patel's plan was to delete genes to change the addresses of different body segments. What happens when you change the pattern of activity in each body segment?

When Patel deleted the *abd-A* gene, the parts of the body that formerly had a *Ubx/abd-A* address now had only *Ubx*. The part

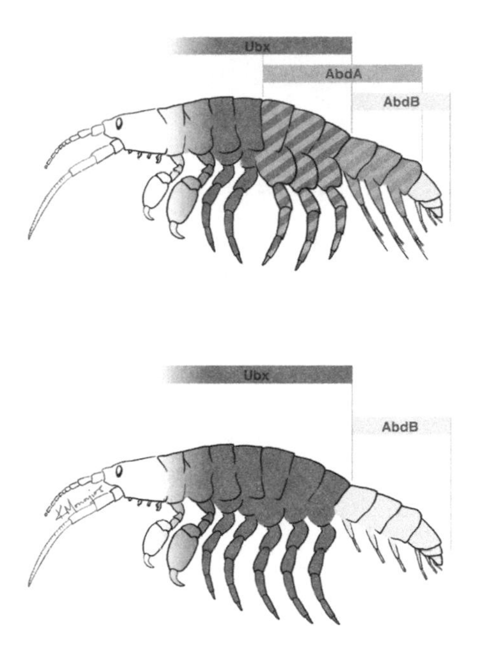

Regular pattern of gene activity (top, shaded areas). Deleting genes to change the patterns of activity in the segments (bottom) changes the kinds of limbs that develop inside.

that had had an *abd-A/Abd-B* now had only an *Abd-B* address. With the change in addresses came a beautiful experimental monster: a creature with backward-facing limbs where forward-facing ones should have been, and stubby ones where bushy ones would normally be. Shifting the patterns of gene activity in the body segments changed which appendage formed in each segment.

Patel found he could change genetic addresses and move appendages around the body at will. In doing so, he wasn't just creating monsters; he was mimicking the diversity of life in nature.

Compare amphipods to their cousins the isopods. Most of us know isopods from one of their most common species: pill bugs. As the name isopod (Greek: "same legs") implies, they have only forward-facing legs, unlike amphipods, which have both forward- and rear-facing legs. When Patel deleted the *abd-A* gene in the amphipod, he made creatures that looked like isopods: they had only forward-facing limbs. He copied nature as well: isopods lack *abd-A* in their normal development.

Changes to these genes explain the differences between creatures as distinct as lobsters and centipedes. The combination of genes active where a lobster's big claw is made is different from those that make a leg. And in creatures like centipedes, where each segment has the same kind of leg, similar genes are active in each body segment. In insects, worms, and flies, these genes form a road map to the body.

The Monster Within

Parhyale, lobsters, and flies are only the start of the story. Frogs, mice, and people have versions of these genes too. They have different names in people and other mammals. Instead of names like *abd-A*, *Abd-B*, and others, they are called *Hox* genes, followed by a number, such as *Hox1*, *Hox2*, and so on. Also, where flies, worms, and insects have only a single string of these genes on one chromosome, we have four sets of these strings on four different chromosomes.

These genes are active along the axis of the bodies of mice and people, and much like flies and *Parhyale*, they are active in different body segments. Our body segments don't sprout wings, or legs that face in every direction. Ours hold vertebrae and ribs. Despite these differences, the question becomes: Does our development work the way it does in *Parhyale* and flies? If you changed the activity of the genes in development, could you make mutants with different numbers of ribs and vertebrae?

Mammalian backbones follow a formula that rarely changes: seven neck vertebrae, followed by twelve thoracic vertebrae, each with a rib, then five lumbar vertebrae. This set is followed by the sacrum and the tail, which in humans is retained as a set of small fused vertebrae called the coccyx.

Just as in flies and *Parhyale*, our different body segments have different addresses of gene activity. For example, one combination of *Bithorax*-like genes marks our cervical region, another the thoracic. Likewise, the boundaries between the thoracic and lumbar regions and between the lumbar and sacral vertebrae both have different genes active inside.

What happens when one genetic address is changed into another? Making mutants is far more difficult in mice than in flies or *Parhyale*. It can take years, largely because the generation time is longer and there are more genes to mutate. But the results are worth the wait.

Take the situation for the lumbar and sacral vertebrae. The region that becomes the lumbar vertebrae has activity of a gene known as *Hox10*. It is followed by the sacral region, which has a genetic address of two genes, *Hox10* and *Hox11*. In a mutant in which the *Hox11* genes are deleted, the segments that would normally form the sacrum have the lumbar genetic address. What happens to the body segments? The end result is a mouse in which the entire sacrum has been transformed into lumbar vertebrae.

Further experiments show this pattern can be repeated with different genes and body parts. Thoracic vertebrae carry ribs. By deleting genes, the entire rear end of the vertebral column can be given the genetic address of thoracic vertebrae. The result: mice with ribs that extend all the way to the tail. As Patel did with *Parhyale*, modifying the genes changes body segments and the organs that develop inside them.

One could call the products of these experiments monsters, but that would hide how beautifully they reveal the mechanisms

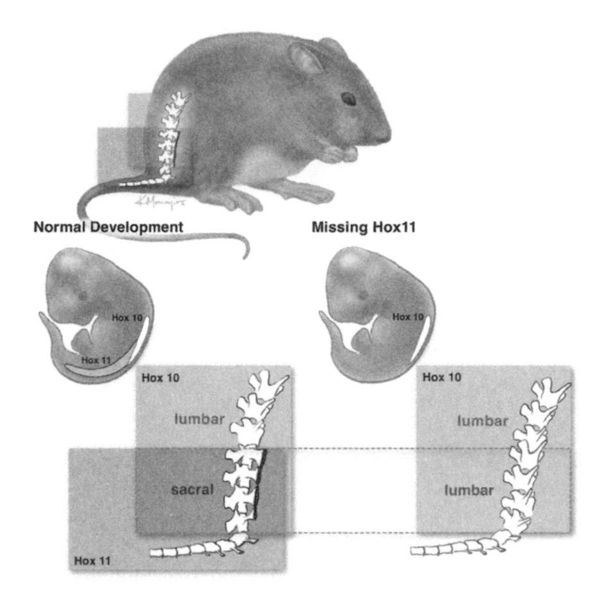

Changes to the activity of *Hox* genes can predictably change sacral vertebrae into lumbar vertebrae.

behind life's diversity. A nineteenth-century observation of life, a discovery in the Fly Room, and modern-day genomic biology combine to reveal beauty inside animal bodies. The genetic architecture that builds the bodies of flies, mice, and people reveals that we are all variations on a theme. From a common toolkit come the many branches of the tree of life.

Reuse, Recycle, Repurpose

As the ubiquity of Lewis's genes in different species was revealed, long-forgotten arcane monographs from the nineteenth century came under renewed scrutiny. In the early 1990s, the observations and ideas of classical natural philosophers such as William

Bateson were fodder for cutting-edge experiments. Bateson had observed that some of the most common kinds of variation entailed changing the number of body parts or having body parts sprout in odd places. Calvin Bridges, Edward Lewis, and the molecular biologists who came later were following a path that had been set nearly a century before. And just as in the nineteenth century, monsters and mutants, whether they were made in the lab or found in the wild, were at the center of it all.

My training was in a world of fossils, museum collections, and expeditions. But one result sent me scurrying to learn molecular biology as quickly as I could.

As teams of researchers around the world explored the activity of *Hox* genes in mice, they found something completely unexpected. Mouse *Hox* genes do not merely control the formation of the vertebrae and ribs along the body axis; they are active in different organs of the embryo, from the head and limbs to the guts and genitalia. It is almost as if these genes are redeployed across the body to build any organ that has its own segmented structure. The patterns of gene activity were pointing to a kind of biological cut-and-paste: a genetic process used to form the main body axis was redeployed to make other bodily structures.

A number of experiments in the early 1990s revealed that the activity of these genes in the limbs is much like that in the body axis; they are active at different times in development and appear to provide a genetic address to different parts of the limb. All limbs, from frogs' legs to whales' flippers, have a similar skeletal pattern. Each has a single bone at the base, the humerus. Then two bones, the radius and the ulna, extend from the elbow. At the end are the bones of the wrist and digits. While the sizes, shapes, and numbers of bones may differ in creatures that use wings to fly, flippers to swim, or hands to play piano, this one bone–two bones–little bones–digits pattern is always there. It is a grand anatomical theme, an ancient pattern that underlies the diversity of every creature with a limb skeleton.

What's more, these three anatomical regions—upper arm, forearm, and hand—correspond to three zones where different *Hox* genes are active. Each region corresponds to a different address of gene activity, much as in the body of a fly, *Parhyale*, or a mouse.

Now researchers could ask, What happens when you change the pattern of genetic activity in the different segments of limbs? We saw in *Parhyale*, and in the body axis of mice, that changing the pattern of gene activity of different body segments could have predictable effects on the organs that develop from them.

In the 1990s a French team of scientists made mutants by deleting *Hox* genes in mice, much as Patel had done with *Parhyale*. When they deleted the *Hox* genes active in the tail, they made a mutant mouse lacking a tail. But now they did the same experiment in the limb. The same *Hox* genes that make the tail are also active in the limb. They define the most terminal segment of the limb—the hand or the foot. When the French team deleted those same genes active in limbs, they made a population of mice with only the one bone–two bones skeleton in their limbs. The mice that developed with the missing genes lacked hands.

I've spent most of my career looking at how hands and feet came about from the fins of fish. My colleagues and I spent six years studying the fossil record to find a fish with arm bones and wrists. Here, suddenly, we had evidence showing the genes that were necessary to make hands.

This result led me to pursue a new path in my own research. In addition to collecting fossils, I realized that I needed to be able to do experiments on genes. Having that toolkit would give me the ability to ask new kinds of questions. Did fish have these genes? If so, what were they doing in fish fins? Could these hand genes help explain how fins were transformed into limbs?

Fish you see at the market, on a dive, or in an aquarium do not have fingers and toes; the fin is made up mostly of a large set of rays with webbing between. The bone in the fin rays is different from the bone of digits. Digits initially form from cartilage precursors, while fin rays develop directly underneath the skin. As we know from the fossil record, the transition from fins to limbs involved two big changes: a gain of digits and a loss of fin rays.

Because the French team revealed the genes that were necessary to make the hands and feet of mice, you might think that those genes are unique to creatures with limbs. But that would be wrong. Fish have these genes too. What are the genes that make hands and feet doing in the fins of fish?

Two young biologists spent four years exploring this question in my Chicago laboratory. First Tetsuya Nakamura worked to duplicate mammalian gene experiments with fish fins. He diligently removed the genes, but the animals lacking these genes did not easily thrive. Remember, these genes are also active in making vertebrae, so the mutant animals could not easily swim. After three years of making mutants, and helping them thrive, Nakamura found something remarkable: when these genes were deleted from the genome, the mutant fish were missing the fin rays.

I first met the second young scientist in 1983, when my anatomy professor, Lee Gehrke, brought his brand-new infant son to a lecture. Little did I know that two decades later the baby, Andrew Gehrke, was going to end up doing a Ph.D. in my laboratory. Gehrke, like Nakamura, would be in the lab until three a.m. most nights devising experiments. A lab in Canada showed that when you marked the hand genes in mice and traced their development, almost all the cells ended up in the wrist and fingers. No big surprise there. The surprise was in fish fins. One late night Gehrke traced the activity of these genes in fish fins and snapped a picture. The resulting figure made the front page of *The New York Times* for the simple reason that it told a big story. The genes that are necessary to build the hands of mice and people are not only present in fish, but they make the bones that sit at the end of the fin skeleton, the fin rays.

The transformation of fins to limbs is a world of repurposing at every level: genes that make hands and feet are present in fish, making the terminal end of their fins, and versions of these same genes help build the terminal end of the bodies of flies and other animals. Great revolutions in life do not necessarily involve the wholesale invention of new genes, organs, or ways of life. Using

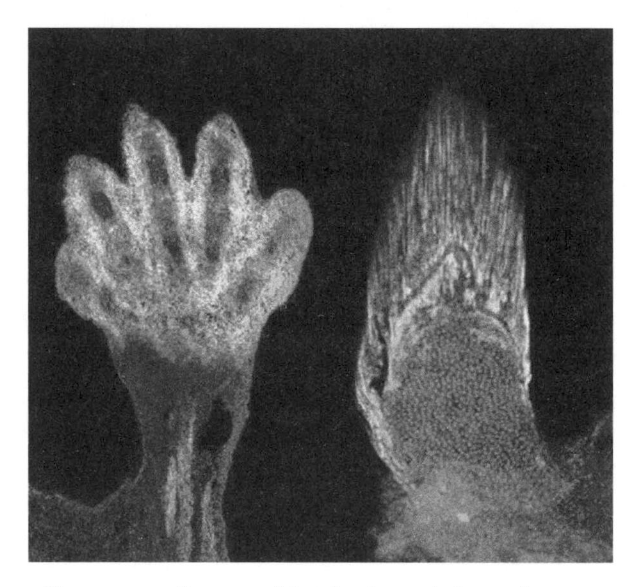

The pattern of gene activity that is needed to make hands (left) is present in fish making the terminal end of their fins. The light area shows where similar *Hox* genes were active during development.

ancient features in new ways opens up a world of possibility for descendants.

Modifying, redeploying, or co-opting ancient genes provides fuel for evolutionary change. Genetic recipes do not need to arise from scratch to make new organs in bodies. Existing genes and networks of them can be pulled off the shelf and modified to make remarkably new things. Using the old to make the new extends to every level of the history of life—even to the invention of new genes themselves.

5

Copycats

IN THE SEVENTEENTH and eighteenth centuries, animal bodies were a frontier every bit as awe-inspiring as expeditions to far reaches of the planet. Basic anatomical features had yet to be discovered in humans, let alone in the diverse creatures collected in remote parts of the Earth. Like peaks, rivers, and other structures in geography, parts of the body were often named after the people who discovered them. Their names connect us historically to the hundreds of greats who explored the structure of bodies for the first time. There is Bachmann's bundle, an electrical tract in the heart. In the eye, there is the annular tendon of Zinn, a ring of fibrous tissue around the optic nerve. And who can forget the Mobile Wad of Henry, which sounds more like a sophomoric joke than the name for the mass of muscles on the external side of the forearm.

The discoverers who coined these names were not just placing their flag on different body parts; they were seeing deep patterns in nature. The French physician Félix Vicq d'Azyr (1748–94) has two structures named after him: the band of Vicq d'Azyr and the bundle of Vicq d'Azyr, both in the brain. As a founder of modern neuroanatomy and later of comparative anatomy, he is an underappreciated figure in the history of science. Vicq d'Azyr was one of the first to compare the anatomical structures of different animals with an eye toward deciphering underlying rules as to why bodily structures look the way they do.

Vicq d'Azyr not only compared similar anatomical structures among species, he looked for organization inside bodies. In dissecting human limbs, he saw that forelimbs and hind limbs were essentially copies of each other. The bones of the arm and leg follow the similar one bone–two bones–many bones–digits array. He extended these comparisons even deeper, seeing how the muscles of the arm and leg follow similar patterns, almost as if they were part of a repeated series of duplicated organs.

Nearly seventy years later, the British anatomist Sir Richard Owen (1804–92) expanded Vicq d'Azyr's idea to the entire body and all animal skeletons. Ribs, vertebrae, and limb bones appear to be modified copies of one another, similar in overall design but with subtle differences in shape, size, and position in the body. Owen was so impressed by this notion that he proposed that the archetype for all skeletons, from fish to people, was a simple creature with blocks of vertebrae and ribs running from tail to head.

Vicq d'Azyr and Owen weren't just uncovering a fundamental pattern in bodies. They were revealing a fact about all of biology itself, most importantly about DNA.

Bridges Again

The careful anatomical dissections of the 1700s and 1800s were a prelude to the painstaking activities in Morgan's Fly Room. In 1913 one of Morgan's students, Sabra Cobey Tice, found a single male fly with extremely tiny eyes. This mutant was rare, the only one among hundreds of normal progeny. By sustaining the flies in the lab and spending a few months finding both males and females, Tice was eventually able to breed more of them.

In 1936, two years before his death, Calvin Bridges decided to use new ultrafine techniques to look at the genetic material of the small-eyed mutants. The technique fit Bridges's precision skills well. He began by removing small patches of cells from the salivary gland, heating them up, placing them on a glass slide, then putting them under a microscope at high magnification to see inside the cells. Doing this right makes the chromosomes visible inside the cells. Bridges didn't know about DNA, but he knew that chromosomes contained genes.

Animal and plant chromosomes come in many different numbers, shapes, and sizes. As we saw with *Bithorax*, when chromosomes are prepared with the techniques that Bridges used, they appear banded, with dark and light stripes, some thick, others thin, alternating in what at first glance appear to be random patterns. The organization of the stripes is the key—they serve as a coordinate system for the position of the genes that Morgan and his team were identifying. Recall that genes are stretches of DNA folded and coiled in on itself to make chromosomes. Sites for genes were identified by where they sat in the rhythm of dark and light bands. A mutation would be revealed by a local change in the pattern of stripes. We now know that the bands are like a GPS with poor satellite coverage; they give a location of the genetic defect of a mutant, but not a precise one.

Bridges prepared the chromosomes of the small-eyed fly mutant and then compared the pattern of stripes to that of normal flies. The stripe pattern was identical except in one region. The small-eyed mutant had a single chromosome that was extra-long, and one whole segment of light and dark bands seemed to repeat the one right next to it. Convinced that this reflected a duplication of one segment of the genome, Bridges made detailed notes and speculated that some aberrant kind of gene copying was the cause for the fly having abnormally small eyes and a longer chromosome.

While Vicq d'Azyr, Owen, and their contemporaries had envisioned bodies as being composed of repeated parts, Calvin Bridges was starting to see copies in the genome. The idea of genetic duplication was just getting going.

Music to Our Genes

Steve Jobs once said, "Picasso had a saying—'good artists copy; great artists steal'—and we [at Apple] have always been shameless about stealing great ideas." What works for art and technology also works for genes. Why build from scratch when you can copy or even steal?

Decades before Jobs uttered these words, a quiet researcher, working mostly alone, was applying them to genetics. Susumu Ohno (1928–2000), at the City of Hope in California, made a hobby of translating the structure of proteins into concert pieces for violin and piano. Knowing that proteins are composed of strings of amino acids, he would use each molecule as a different note. The music had a deep, almost mystical resonance for him. The score made from a malignant cancer-causing protein sounded to him like Chopin's Funeral March. A score made from the sequence of a protein that helped the body process sugars was, to his ear, a lullaby. Ohno found more than dirges and melodies in genes and proteins—he found a new view of biological invention.

Ohno had been raised by the minister of education of the Japanese viceroyship in Korea and was fortunate to have educa-

Susumu Ohno (left)

tional opportunities and intellectual challenges from a very early age. By his own account, his life's work came from his childhood love of horses. Spending weekends riding, he came to the opinion that "when a horse is no good, there is not much you can do." To Ohno, the key to understanding different horses lay in understanding the genes that made them faster or slower, stronger or weaker, bigger or smaller. Pursuing genetics both in Japan and later at UCLA, he was familiar with Morgan and Bridges's work and spent his days studying chromosomes for patterns that would describe the similarities and differences among living things.

In the 1960s, using techniques not much different from those of Bridges decades before, Ohno stained the cells of different mammal species with chemicals to reveal the bands of their chromosomes. Then he took pictures of them, cut them out like paper dolls, and laid them on a table. With the cutout photos of the chromosomes in front of him, he was able to ask, What are the differences among the chromosomes of diverse species? It was an ingenious and low-tech approach to get at the genetic changes that make species different.

Ohno started by comparing the chromosomes of species of mammals, from tiny shrews to giraffes. After he obtained cells of different species from zoos and other sources, his first observation was that the total number of chromosomes of different species can vary widely, from a low of seventeen pairs in a creeping vole to eighty-four pairs in the black rhinoceros.

Ohno then did something elegant in its simplicity but powerful in its implications. He weighed the paper cutouts of the chromosomes for each species. He surmised that the weight of the cutouts could serve as a proxy for the total amount of genetic material that was inside a creature's cells. He was weighing cardboard cutouts of pictures of chromosomes, not the chromosomes themselves, but it was the relative weights that mattered. For this to work, Ohno had to cut the chromosomes from the pictures very carefully. When he weighed the cutouts of the seventeen chromosomes of the vole and the cutouts of the eighty-four black rhino chromosomes, the total weight for each species was virtually the same. In fact, the cutouts of all the different species of mammals weighed the same, from elephants to shrews. Ohno concluded that the similar weights of the cardboard cutouts showed that the weights of the chromosomes were the same in different mammals. This similarity held true despite large differences in the number of chromosomes in the various species.

Ohno extended his comparison to other creatures: Did different species of amphibians and fish also have the same amount of genetic material? Species of salamanders tend to look alike, and Ohno assumed that their genetic material should be virtually the same. Cutting out the chromosomes and weighing them brought the big surprise: different, but anatomically similar, species of salamanders might have widely varying amounts of DNA in their cells, with some species having five to ten times as much as others. The same was true for species of frogs. What's more, the amount of genetic material of both kinds of amphibians dwarfed that of humans and other mammals. Some salamanders and frogs have twenty-five times more genetic material than humans.

With his cardboard cutouts, Ohno discovered something that billions of dollars of genome projects were to confirm decades later. The complexity of an animal and the differences among species do not correspond to the amount of genetic material in cells. Because salamanders generally looked alike despite one species having ten times more DNA than another, and that extra genetic material did not seem to be related to any observable difference in the animals' anatomy, Ohno surmised that the genomes of salamanders and other species are riddled with meaningless stretches of DNA. This DNA was, to use his term, "junk."

Ohno noticed that salamanders with the largest genomes tended also to have strange banding patterns along their chromosomes: entire stretches seemed to be made up of repeated or duplicated bands. He speculated that all the extra DNA in the cells of salamanders and frogs came about because of duplicated genes, as if some parts of the genome had been copied over and over and over again. All that "junk" came from a copying process gone wild. Ohno suspected that duplication run amok was a major factor in the great transitions in the history of life. Like a good detective, he sought to understand how this happened and what it might imply about the evolutionary past.

Ohno knew that when cells divide, chromosomes are copied and mistakes can happen. T. H. Morgan's group in the Fly Room had watched cells divide. By banding the chromosomes, they had seen how they copied and the kinds of errors that happened within cells. Most animals have two sets of chromosomes in each cell, one from each parent. Humans have twenty-three pairs of chromosomes, each pair containing one chromosome from the mother and one from the father, giving us a total of forty-six chromosomes. While most of our cells have two copies of each chromosome, the sperm and the egg have only one. When sperm and eggs are manufactured, the DNA replicates and chromosomes get copied, and only one set of chromosomes gets allocated to each sperm and egg. But things can go wrong. When the chromosomes are copied, the new pairs can often trade material. If the exchange is unequal, one chromosome can end up with extra copies of genes, the other fewer. This process could produce offspring with many copies of the same gene and a larger genome as a result, much like what Bridges saw with his small-eyed fly, or Ohno with his cardboard cutouts.

Another type of error can change the entire genome. After chromosomes get copied, they move to new sperm and egg cells. If they don't move correctly to their new homes, some sperm or eggs can end up with extra chromosomes. This is a duplication not just of a single gene but of the many thousands that can lie on the chromosome. The sperm or egg can now make an embryo not with the normal two sets but sometimes with a single extra straggler of a chromosome or whole sets of them. Instead of two copies of each chromosome, the sperm or egg might end up with three or more.

The presence of a single extra chromosome can bring about dramatic changes. Often, with the balance of genetic material altered, the fine interaction of genes necessary for normal development is disrupted. One result can be a birth anomaly. Down syndrome comes about when the embryo ends up with an extra copy of chromosome 21. The syndrome affects the entire body, from the nervous system to the chin, eyes, and creases across the palm of the hand. Geneticists have assembled catalogs that describe what happens with chromosomes, from Patau syndrome, where the embryo has an extra copy of chromosome 13, to Edwards syndrome, which results from an additional chromosome 18. In both conditions, the development of the brain, skeleton, and organs—virtually every part of the body—is affected.

It is one thing to have a single extra chromosome; it is something else altogether for an embryo to end up with duplicated sets of them. Biological magic can happen. Instead of the normal two copies of each gene, it might have three, four, or even sixteen or more. At almost every meal, we consume individuals with extra sets of chromosomes. Bananas and watermelons have three sets; potatoes, leeks, and peanuts have four; strawberries as many as eight. Plant breeders realized early on that by breeding plants with entirely duplicated genomes, the offspring will sometimes have extra sets of chromosomes and be more vibrant or tastier. Nobody knows why, but some think the extra genetic material is put to new uses to make growth and metabolism more robust.

This boost of chromosomes happens regularly in nature. When a sperm with an extra set of chromosomes fertilizes an egg with an extra set, the embryo can be viable, even more robust. This new individual will be different from its peers. On occasion, because its genome is so different from that of its parents or brethren, it can breed effectively only with individuals that also have the extra set of chromosomes. They are a kind of hopeful monster, a genetic mutation produced in a single step, by a change in the allocation of chromosomes to sperm and egg.

There are over six hundred thousand species of flowering plants in the world. More than half of them have duplicate sets of chromosomes, their species formed by a simple shift in how sperm and egg are made.

What is common in plants is rare in animals. Such mutants are rarely viable in mammals, birds, or reptiles. The animals that have significant numbers of species with extra sets of chromosomes are reptiles, amphibians, and fish. Lizards can often be born with multiple sets of chromosomes. Individuals that have this condition grow and look normal but are typically sterile. Frogs and species of fish, however, can have multiple sets and breed normally.

When Ohno made his cardboard cutouts, he knew that simple errors in the cell could duplicate chromosomes, parts of chromosomes, even entire sets of chromosomes. He therefore envisioned a world of copies and copies of copies. To him, duplicates were the seeds of invention.

Salamander and frog cutouts inspired a new view of genetic inventions in the history of life. A prevailing idea was that the fuel for evolution by natural selection was small changes in genes. What if, Onho postulated, an engine for evolutionary change was gene duplication? Inventions would come ready-made for new uses. If a gene gets duplicated, two genes now exist where there was once only one. This kind of redundancy means that one gene can stay the same and preserve the old function, while the other copy can change and gain a new one. A new gene can be produced in a whoosh at almost no cost to the bearer.

Duplication can set the basis for change at every level of the genome. Useful parts come ready-made to take change in new directions—using the old to make the new.

By the time Ohno finished making his cutouts of chromosomes, the sequences of different proteins were becoming available. They only confirmed the extent of the copying that happened in the genome. It was copies all the way down: whole genomes could be copied, genes could be duplicated, even parts of proteins seemed to have repeated sequences inside. These duplicated proteins, to Ohno, made special music. Ohno and his wife, Midori, a singer, were often called upon to perform some of their music of duplicated molecules at social events.

Copies Everywhere

The genome at every level resembles a musical score in which the same musical phrases are repeated in different ways to make vastly different songs. In fact, if nature were a composer, she would be one of the greatest copyright violators in history—everything, from parts of DNA to entire genes and proteins, is a modified copy of something else. Observing duplications in the genome is like wearing a new pair of glasses: the world looks different. Once you see duplications in the genome, you see them everywhere. New genetic material looks like copies of old stuff that was repurposed for new uses. The creative power of evolution is more like a copycat who duplicates and modifies ancient DNA, proteins, and even the blueprints that build organs, for billions of years.

The first people to look at protein sequences, including Zuckerkandl and Pauling, ran right up against duplications. Hemoglobin, the protein that transports oxygen in blood, exists in many forms, each corresponding to a different condition of life. The needs of a fetus differ from those of an adult. In the womb, oxygen comes from the mother's bloodstream, whereas in adults, lungs are involved. These life stages are marked by different hemoglobins that are copies of each other.

Different amino acid sequences of proteins seemed to be versions of one another. You can find examples in every tissue and organ—skin, blood, eyes, and noses, to name a few.

Keratin is a protein that gives our nails, skin, and hair their special physical properties. Each tissue has a different kind of keratin inside, some pliant, some hard. The keratin gene family formed as a single ancient keratin gene that was duplicated to make keratins dedicated for each tissue.

Color vision happens through the action of proteins called opsins. People see a wide range of colors because we have three opsins, each tuned to a different wavelength of light: red, green, and blue. These opsins have undergone duplications from a single one to the full set of three, with an expansion of visual acuity.

A similar pattern holds for the molecules that aid in smelling. The repertoire of smells that an animal can perceive is in large part defined by the number of olfactory receptor genes it has. Humans have about five hundred of them, but we're nothing compared to dogs and rats, which have a thousand and fifteen hundred, respectively. (Fish have about 150.) For vision, olfaction, breathing, and virtually everything else animals do, duplicate genes make it all happen. Almost every protein in the body is a modified duplicate of an ancient one, repurposed for new functions.

As Lewis and others who followed him saw, genes that build bodies are often modified copies of one another. Lewis's genes, *Bithorax* in flies, and *Hox* genes in mice are duplicates. *Hox* genes, so involved with body architecture, are a large gene family that has, over time, only increased in number. Humans, like mice, have thirty-nine, whereas flies have only eight. The same is true for other major toolkit genes that build animal bodies. Genes of the *Pax* family play a role in the formation of eyes, ears, the spinal cord, and internal organs. There are nine of them. *Pax 6* is involved in eye development, *Pax 4* in the pancreas. Embryos lacking these genes do not have these organs. Their grandparent gene was a single *Pax* gene that got duplicated, with the different copies gaining new functions in different tissues and organs.

We now know that genes in the genome are part of gene families, filled with duplicates, that share essential sequences. A family can consist of a handful of genes or thousands of them, each with different functions. And these speak about a powerful process at work during evolution.

As Ohno saw, copies can be paths to invention. My Chicago colleague Manyuan Long looked at fruit flies to estimate how new genes came about in different species. Long made use of genome sequences that were available for different species of fly. More than five hundred new genes differed between the species, about 4 percent of the whole genome. While some came about from processes we don't yet understand, most of the new genes arose as duplicates of ancient ones. Why invent from scratch when you can copy?

Gene duplication can even get personal.

Big Brains

A signature human trait is our enlarged brain relative to our primate relatives. Obviously, knowing the genetic basis for its origin would tell us how thinking, talking, and many of our other unique abilities arose. Judging from the fossil record, the volume of the brain has nearly tripled in size from that of our australopithecene ancestors three million years ago. The brain expanded in particular regions, most notably the so-called cortical region of the forebrain, associated with thoughts, planning, and learning.

The fossil record shows that the expansion of the brain was related to other changes, most notably a new complexity in the kinds of tools that our ancestors made and used. Now onto the scene comes genomic technology, opening up a new quest: understanding the genes that make us human.

One approach would be to compare the genomes of humans and chimps. You'd end up with a list of genes that humans have but chimps do not. While that list would be informative, it would not say anything about which genes are important for the origin of the human brain. The differences could relate to any feature that separates humans from other primates or even none at all.

One way into this problem sounds like it comes from science fiction: grow brains in a dish. Even the name, organoid, has that ring to it. The idea is to take brain cells from a developing animal, put them in a dish, and see under what conditions brain structures can be made. It is far easier to study tissues in a dish than in the embryo, particularly in mammals, where most of the action happens in the womb.

A team in California compared brain organoids of humans and rhesus macaques and made lists of all the differences. In the dish, a version of the uniquely human cortical region formed in the human organoid but not in that of the monkey. The researchers looked at the genes that were turned on when this tissue was forming. One gene was active in every human cell but lacking in monkey tissue. The name, *NOTCH2NL*, is a mouthful but is relevant to the story.

At the same time, a lab six thousand miles away in Holland had unusual access to human fetal brain tissue from miscarriages and medically necessary abortions. This tissue was unique, coming from embryos at the stage when the brain was forming. The researchers probed the genes that were active in the brain and found a small number that had the right profile to be brain-forming—they were turned on at the right time in development and were actively making proteins. One of them was *NOTCH2NL*, the gene identified in the dish experiments.

The science fiction flavor of the research only increased when the Dutch team took the human *NOTCH2NL* and inserted it into a mouse. They made a human-mouse chimera. The result was a mouse that grew more cortical brain cells, much like a human.

The California team then looked at the genome, comparing that of humans, Neanderthals, and primates. They found that the *NOTCH2NL* gene was one of three at work in human brains, and all of them were similar to a single gene, *NOTCH*, which is present in everything from flies to primates and is involved in the development of many different organs. How did the three human brain genes originate? By duplications of the primitive *NOTCH* gene from primate ancestors. Once they were duplicated, the copies gained new functions.

Gene duplications not only help explain the past, they factor into the present day. The three *NOTCH* duplicates sit end to end in the human genome. This structure makes the region unstable, able to break when the genes are copied during cell division. The breaks are places where the chromosome can be damaged. These changes affect the function of the genes and of the brain. When the cells divide, the region can be duplicated or deleted. The ones with the duplications grow up to have larger brains; those with the deletions have smaller ones. While some individuals with these genetic changes have normal brain function, most show symptoms of schizophrenia and autism.

Clearly *NOTCH2NL* is not the only gene involved in making large brains. But as this work shows, our genome is chock full of repeats, gene families, and other kinds of copies, and these duplications can be fuel for invention and change.

Copies Gone Wild

Roy Britten had science in his DNA. Born in 1912 and raised by parents in different scientific disciplines, he took to physics, ultimately landing a job with the Manhattan Project during World War II. With each passing year, his pacifism increased, and he yearned for a new job. Eventually he found one, working for a geophysics lab in Washington, D.C. After the discovery of the structure of DNA in 1953, and always seeking new intellectual adventures, Britten took a short course on viruses at Cold Spring Harbor Laboratory in New York. Armed with that knowledge and seeing DNA as a new frontier, he set out to work on its structure.

The problems that consumed Britten involved understanding how many genes there are in the genome and how they are organized. These were the days before genomes could be sequenced, and its organization was mostly a mystery. Lacking gene sequencers, Britten, like Ohno before him, had to conjure some clever experimental tricks.

Following Ohno, Britten had a hunch that the genome was composed of duplicated parts. He designed a clever experiment to approximate how much of the genome contains copies. He removed DNA from a creature's cells, then heated it, breaking the DNA strand into thousands of smaller pieces. Changing the conditions, he let the strand come back together. The trick was to measure how fast the different parts came together into another single strand. He surmised that the speed at which the DNA reassembled would give him a sense of how many repeated elements were in the genome. The reason? Due to the chemistry of the DNA molecule, "like finds like" more quickly than otherwise. A genome composed of repeated parts, which are alike, should come back together more quickly than one composed of fewer repeated ones.

Britten did his first calculations on the DNA of a calf and a salmon, then expanded the comparison to other species. Even though he expected to find lots of duplicates in the genome, he was shocked by his results. By his estimates, about 40 percent of the genome of the calf was made up of repeating sequences. In the salmon, the number came closer to 50 percent. The sheer number of repeats in each genome was as surprising as their prevalence in different species. Almost every animal's DNA that he broke apart and reassembled had an enormous number of repeated elements inside. Using the crude techniques available at the time, he estimated that some elements had over one million copies in the genome.

The advent of genome projects means that we can see the specific sequences that have been duplicated in the genome and give finer resolution to the early efforts of Bridges, Ohno, and Britten. A fragment called *ALU*, about three hundred bases long, is seen in all primates. Fully 13 percent of the human genome is composed of *ALU* repeats. Another short fragment, *LINE1*, is repeated hundreds of thousands of times in the human genome and makes up 17 percent of it. All told, over two-thirds of our entire genome is composed of strings of repeated copies of sequences with no known function. Duplication in the genome has run amok.

Roy Britten published scientific papers into his nineties, until his death from pancreatic cancer in 2012. One year before his death, he published a paper with new findings in the *Proceedings of the National Academy of Sciences* with a title that would have made Ohno smile: "Almost All Human Genes Arose by Duplication."

Corny Genes

Barbara McClintock (1902–92) launched her career wanting to follow in T. H. Morgan's footsteps to understand the basis of genetics. Unfortunately, when McClintock entered Cornell University, women were not allowed to major in genetics, so she enrolled in an approved "ladies' major," horticulture. But McClintock got the last laugh. She ended up joining a team that broke new ground studying the genetics of corn.

As a subject of study, corn had a distinct advantage over Morgan's flies. A single ear of corn can have as many as twelve hundred kernels. McClintock knew that they were ideal for genetics study because each kernel is a separate embryo, a distinct indi-

Barbara McClintock with corn

vidual. Next time you eat an ear of corn, imagine that you are eating over one thousand genetically distinct creatures. For McClintock, each ear of corn became a nursery in which she could explore genetics. What's more, corn comes in many varieties, with kernels of different colors ranging from white to blue to speckled. One ear of corn could be the basis for an experiment tracing thousands of individuals. Experiments could be fast, cheap, and rich with data.

McClintock started her work much as Morgan's team had, by developing techniques to visualize chromosomes. Treating corn with a number of stains, she was able to map regions of them in great detail with light and dark bands. Then she got lucky. She found a region of the corn chromosome where the chromosomes would simply break apart, as if there were some structural defect at that particular spot. Homing in on it, she mapped that region in great detail in different kernels of corn. To her surprise, she found that the break point hopped around the genome. This single insight was one of the great ideas in the history of genetics: the genome is not static—genes can jump from place to place.

McClintock did not stop there. A careful and thorough researcher, she held off on telling the world of this discovery until she traced its implications. She asked, Did the jumping genes have any effect on the kernels themselves? What if a jumping gene landed on the site of another gene?

McClintock used special properties of corn kernels to find the answer. The outer pigment develops as cells multiply. It starts with a single cell that continually divides. If that starter cell is a particular color, say purple, the entire kernel will be made of its descendant cells, all of them purple. But imagine that a genetic change happens to one cell during that process, so that the purple gene acquires a mutation. The daughter cells of that particular cell won't be purple, they'll be the default color, usually white. That white cell will continue to divide to produce a batch of white cells. The end result will be a mostly purple kernel with a splotch of white.

By tracing the different patches of colors in each kernel, McClintock could see where and when mutations were happening in the genes inside. She could look at mutations on each kernel and repeat this with thousands of them on every cob. McClintock studied hundreds of thousands of kernels, breeding corn to make different colors with different kinds of patches. She found that mutations in the colors can be switched on and off, then on again. Studying the chromosomes, much as Bridges and Morgan had, she discovered that the mutations happened when the chromosomal break point region jumped and landed inside a pigment gene. When it inserted in a pigment gene, it would corrupt it, and the pigment would no longer be made. When it jumped out, the pigment would be made again. The corn genome was filled with genes that were making copies of themselves, hopping around and in so doing making different color patches.

After spending decades on the work, McClintock presented her idea of jumping genes at a talk at Cold Spring Harbor Lab, where she worked. The collected experts could not have cared less. People didn't understand her, didn't believe her, or thought maybe her discovery was just something weird about corn. McClintock described their reaction by saying, "They thought I was crazy, absolutely mad."

There the problem sat for decades. But McClintock was undeterred, mapping jumping genes in thousands of ears of corn. Her attitude at the time was, "If you know you're right, you don't care. You know that sooner or later, it will come out in the wash."

Then, in 1977, other laboratories found evidence of jumping

genes in bacteria, in mice—indeed in every single species they tested. Another surprise came from looking at the genes themselves. Our genome has been taken over by jumping genes—about 70 percent of it is made up of them. Jumping genes are the rule, not the exception. Those hugely repetitive fragments in our genome, *ALU* and *LINE1*, the ones that are so repeated that they have millions of copies? These are jumping genes that make copies of themselves and insert themselves all over the genome. Roy Britten had been seeing them with his elegant, yet crude, experiments in the 1960s.

McClintock won the Nobel Prize in Physiology or Medicine in 1983 for her discovery. Back in 1970, President Richard Nixon presented her with the National Medal of Science. During the ceremony, Nixon offered a somewhat garbled take on the scientific enterprise, but one that nevertheless recognized her impact: "I have read [explanations of your scientific work,] and I want you to know that I do not understand them." He continued, "But I want you to know, too, that because I do not understand them, I realize how enormously important their contributions are to this nation. That, to me, is the nature of science."

The genome is not a stale and static entity. Genomes are churning with activity. Genes can duplicate, and entire genomes can be copied. Genes can make copies of themselves and jump around the genome.

Imagine two kinds of genes in the genome: some that have a function and make a protein, and others that live just to jump around and make copies of themselves. Over time what will happen? All else being equal, the copiers will occupy ever greater parts of the genome. This is one reason two-thirds of our genome is composed of repeated sequences, like *LINE1* and *ALU.* Unchecked, they will take over. The only thing stopping these parasites is that if they get completely out of control, they could cause the death of their host, and over time they, too, will die. Individuals who carry jumping genes that are completely uncontrolled will die and not pass them on. The selfish genes and their hosts are in tension, even at war with each other, as the selfish genes live just to make copies of themselves and host genomes struggle to contain them.

As with Apple under Steve Jobs, copying is a mother of invention: plagiarism in the genome is the source of countless genetic inventions. Much as in technology, business, and economics, disruption can bring revolution. Animal cells have been undergoing disruption for billions of years, and as we'll see, these changes can bring about whole new ways of life.

6

Our Inner Battlefield

SEEDS FOR MY WORK were sown during a weekly ritual I performed while I was a graduate student in the 1980s. Every Thursday morning I would trudge up five flights of stairs to a large storage area in Harvard's Museum of Comparative Zoology. Home to the bird collection, the space had creaky wooden floors and twenty-foot-high ceilings. The walls were lined with cabinets and shelves filled with skeletons, feathers, and skins collected during expeditions of the nineteenth and twentieth centuries. The smell of the mothballs that protected the skins wafted through the air. History also permeated the place, both for ornithology and for science as a whole. That link to the past was what drew me: my pilgrimages were to meet with the eighty-year-old retired bird curator, Ernst Mayr.

By the mid-1980s, Mayr was among the last living members of a generation of geneticists, paleontologists, and taxonomists who had defined the field of evolutionary biology in the mid-twentieth century. Mayr's part in this scientific achievement was to write one of the classic books of this time, *Animals, Species and Evolution*, an immense tome that guided research for a generation of scientists on the formation of new species.

Each week I'd arrive with a question and share a pot of tea with the great man as he held forth on the history of the field while offering spirited opinions on the ideas and personalities that shaped it. In advance of each visit, I'd scavenge the literature to generate a good subject to serve as fodder for his reminiscences. Transported in time and space by his stories, I felt incredibly fortunate to have such an amazing gig at the start of my own career.

One Thursday I came with a book, *The Material Basis of Evolution*, by the German scientist Richard Goldschmidt, a paperback reprint of a volume first published in 1940. Showing it to Mayr, I saw his face turn beet red as his eyes shot through me with an icy glare. He rose, stood still, and didn't so much as acknowledge my presence for an interval that felt interminable. I had crossed some hidden line and was quite certain that I could say farewell to my Thursday teas.

Mayr walked silently to an old wooden file cabinet and rifled through its contents. He returned with a yellowed reprint of one of Goldschmidt's papers and slapped the article onto the table, saying, "I wrote my book in response to the crap in the first sentence of a paragraph toward the end of this." Taking his cue, I thumbed through the paper until I hit page 96. It was unmistakable; on it were more angry marginalia than original text.

Three and a half decades had passed between the publication of Goldschmidt's article and Mayr's rage. How could a single sentence, let alone idea, evoke such passion and catalyze an 811-page book that itself launched entire research careers?

At issue was how changes in genes could bring about new inventions in the history of life. The conventional view was that inventions emerge gradually over time with small genetic changes at each step. This notion was supported by such a large body of theoretical and laboratory work that it was almost taken as axiomatic. The British statistician Sir Ronald A. Fisher derived it mathematically in the 1920s as he tried to link the emerging field of genetics with Darwinian evolution. Part of the logic is embedded in the idea that if you were to make a random change to a system, large changes are more likely to be bad, often catastrophically so, than smaller ones.

Take, for example, an airplane. Any random change that departs dramatically from the norm is almost certainly going to lead to an airplane that can't fly. Randomly changing the shape of the body; the position, form, or shape of the engines; or the configuration of the wings would likely lead to a grounded monstrosity. But small tweaks, such as to the color of the seats or minor alterations in size, are less likely to be dire. Indeed, they have more of a chance of increasing performance than large changes do, even marginally. This kind of thinking dominated the field of evolutionary biology for years, to the point that challenging it was akin to denying that gravity causes apples to fall from trees.

Goldschmidt, a refugee from Nazi Germany, entered academe in the United States having studied mutants for decades. With his move to North America, he crashed the party in the field of genetics, unconcerned by the status quo. Impressed by mutants with two heads or extra body segments, such as those Calvin Bridges was discovering, he thought that a major transformation could happen in a single step with a single dramatic mutation. The drama behind the idea is captured in one of Goldschmidt's most famous remarks, indeed the one that enraged Mayr so thoroughly: "The first bird hatched from a reptilian egg." No gradual change here—in his view, biological revolutions happened with a single mutation in one generation.

Goldschmidt's mutants were given a name: "hopeful monsters." They were monsters because they differed so dramatically from the norm, and hopeful in that they were the seeds for an entire revolution in the history of life. In the world of plants, where changes in chromosomal numbers could bring about new species all at once, Goldschmidt's idea was not controversial. For animals, however, things were very different.

The assault on Goldschmidt's idea was immediate and fierce. The most salient criticisms challenged the chances that a hopeful monster could be viable and ultimately reproduce. First, the mutation would need to make viable and fertile offspring. It was well known by that time that most mutants, let alone dramatic ones, were either sterile or died before they could give rise to offspring. Even if a mutant were to survive and be fertile, its fate would still be unsure. It wouldn't do if only a single mutant were present in a population—it would need to find a mate that also had the mutation. For Goldschmidt's hopeful monster to give rise to a major revolution in a single step, a chain of unlikely events would have to happen: a major mutation would have to make a viable adult; it would have to happen in males and females simultaneously; and some of those individuals would need to be able to find each other, mate, and rear their own offspring, which themselves could reproduce.

By the time I studied biology in the 1970s, Goldschmidt's reputation remained something between a pariah and a heretic, as somebody who had dared publish a view so obviously wrong. Not only did he publish it, he seemed to relish his contrarian role, spending the final decades of his career defending hopeful monsters, often to public ridicule.

Mayr, Goldschmidt, and their contemporaries were debating one of the central issues of life's diversity—how major evolutionary changes happen. Although Goldschmidt's hopeful monsters were implausible, open questions remained. The issue was not with gradual change; biologists have long known that small incremental genetic changes could lead to massive revolutions over the millions of years of geological time. A deeper puzzle emerges from the fossil record. Take, for example, the origin of a skeleton, one of the biggest events in our own species' history. For millions of years, wormlike ancestors lived with no bones inside their bodies. Bone has a characteristic structure, with highly organized layers of cells that manufacture the distinctive proteins and crystals that give the skeleton its rigidity and regulate its ways of growing. The origin of a skeleton allowed our ancestors to get large and have a rigid body to find prey, avoid predators, and move about. This invention arose because of the emergence of a new kind of cell, one that can produce the proteins needed to make skeletons, nourish them, and help them grow. But different kinds of tissues—whether skin, nerve, or bone—are made by cells that make hundreds of different proteins. Nerve cells are distinct from skeletal cells because numerous proteins give them the ability to conduct nervous impulses. These, of course, are lacking in the skeleton and the cells that build it. Likewise, cartilage, tendon, and bone are made from proteins that nerve cells do not produce. And the skeleton is only one example: the nearly 600-million-year history of animal life involved the origin of hundreds of new tissues, which enabled new ways of feeding, digesting, moving, and reproducing.

And here is the challenge: the origin of new tissues and cells from those of ancestors requires changes to hundreds of genes. How could new cells and tissues arise if a multitude of separate mutations must happen simultaneously throughout the genome? If the odds of one incremental mutation happening are relatively small, then imagine the impossibility of hundreds of them happening at once. This would be akin to winning the jackpot on not just one roulette wheel but every single wheel in a casino at the same time.

Pregnant with Meaning

It is hard to miss my University of Chicago colleague Vinny Lynch in the gym: sporting tattoos of a menagerie of species on his arms and legs, he stands out even among inked college students. Dragonflies and fish in a river scene populate his appendages.

The river scene is an homage to the Hudson River ecosystem that nurtured his childhood love of science. Growing up in a town along its banks, he developed a passion for the creatures that lived at the water's margin. Documenting, drawing, and reading about different animals transported him to another world. Unfortunately, his curiosity about life's diversity did not translate into success in school. He was a failure because, as he described it, he "didn't listen to lectures"; instead he stared out the window at birds and insects.

Fortunately, one biology teacher saw through his idyll and let him sit at the back of the class with books and field guides that she'd quiz him on later. This experience provided by one sage instructor propelled him to a career in biology. He has spent his life ever since exploring how animal diversity comes about: not just how animals live, eat, and move, but how over millions of years they arose from distant ancestors. And his specialty is applying high technology to these deep questions.

Progress in biology is as much about defining the right question as it is about finding an experimental system in which to explore it. T. H. Morgan found clues to genetics in flies. Barbara McClintock came to understand the working of genes in corn. Vinny Lynch is finding clues to the great revolutions in the history of life in decidual stromal cells.

Lynch's eyes widen as he describes decidual stromal cells.

When we first chatted about them, he gushed that they are some of the "most beautiful cells in the body." I'll admit it sounds impossibly nerdy, but once I saw them under the microscope, I came to agree. Most cells look like regular little dots under higher magnification. Not these. With big red bodies and rich connective tissue in between, they look almost lush, if you can apply that term to cells.

For Lynch, the beauty of decidual stromal cells is not only aesthetic but scientific. They are a window into the origin of one of the great inventions in the history of life: pregnancy. Most fish, birds, and reptiles, even very primitive mammals, hatch from eggs. They do not have the mammalian style of pregnancy, where the embryo develops within the mother and shares her blood supply. They also do not have decidual stromal cells.

Pregnancy seems at once completely natural and utterly miraculous. Sperm maneuver through the uterus and fallopian

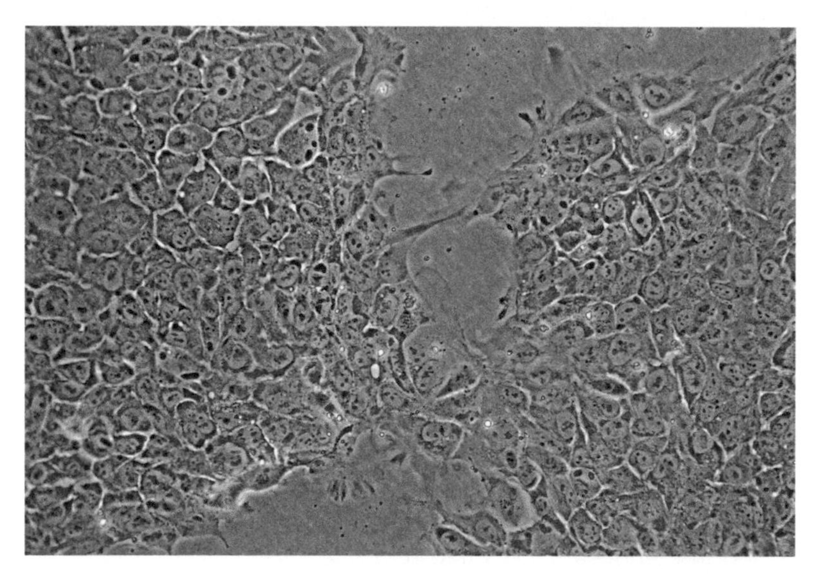

A beautiful cell: decidual stromal cells

tubes to ultimately find the egg. Then one sperm (in rare cases more) enters the egg and sets off a chain reaction of events. Sperm and egg genomes merge, and the two become a single cell. Over time that cell gives rise to a body made of trillions of cells all packed in the right place. A placenta and umbilicus form to connect the mother and the fetus housed in the protective womb. For the womb to hold the fetus, a suite of new structures has to be constructed.

Fertilization brings about a cascade of changes in the body of the mother as well. In the uterus, specialized cells form to connect the fetus to her, bringing their blood supplies in close proximity. These cells mask the fact that the fetus is an alien inside the mother, having a contribution of genes and proteins from the father. There is always the risk that the mother's immune system could go on a search-and-destroy mission for paternal proteins and kill the fetus. Specialized cells dampen those differences. The cell that does much of this magic, from buffering the mother's immune response to channeling nutrients to the fetus, is the decidual stromal cell.

The trigger that makes these cells and initiates many of the changes in the uterus is a spike in the hormone progesterone in the mother's bloodstream. On a monthly basis, progesterone rises in the mother's bloodstream, and the uterus prepares for pregnancy. When progesterone contacts cells of the uterus, it causes them to multiply and change, making the lining of the uterus, the endometrium, thicker. The rising progesterone levels cause a set of cells known as fibroblasts to change into decidual stromal cells. If pregnancy does not happen that month, the cells slough off. But if pregnancy is achieved, the ovaries start to make progesterone, the cells and the rich cellular medium that lines the uterus continue to grow, and the decidual stromal cells form and start to do their work.

Lynch's fascination with these cells derived from a scientific talk he attended in Texas while he was a graduate student at Yale University. A researcher, speaking about pregnancy, showed slides of decidual stromal cells. Lynch learned that these cells had a special property: you could make them in a dish. The researcher had found that when he took normal fibroblasts from anywhere in the body, put them in a petri dish, and added a cocktail of progesterone and some other chemicals, he could make normal decidual stromal cells. Unknown to Lynch at the time, and by sheer coincidence, all this work was being done at Yale in the building next door to his own.

Lynch quickly learned to make decidual stromal cells in the controlled environment of the lab. He now could probe their genomes to see how they had come about millions of years ago. He had at his disposal a very powerful new technology, one that makes use of incredibly fast gene sequencers. Using this technology, he could look at a cell, or an entire tissue, and see the sequence of every single gene that was active in it—all of them at once.

Think about what a technology like this can do. If the differences between cells arise from the genes active in each one, then identifying the constellation of genes turned on in different cells becomes a critical part of the quest to understand what makes cells distinct. Recall that a nerve cell differs from a bone cell because different genes are making different proteins inside each. Likewise, a decidual stromal cell is distinct from a fibroblast in the genes that are active within. Lynch could look at one cell and compare it to another to ask fundamental questions: What are the differences in gene activity between the two cells? Is it one gene that makes them different, or is it several acting together, and if so, which ones are they?

Lynch took fibroblasts, put them in the dish, hit them with

progesterone, and turned them into decidual stromal cells. Then he looked at which genes were activated. The result was as surprising as it was formidable. The origin of decidual stromal cells didn't involve a single gene, or even a handful of them, being activated. Rather, hundreds of genes were turned on at the same time.

Decidual stromal cells are unique to mammals—no other creature has a version of them. Their origin is a central part of the origin of pregnancy itself. But therein lies the problem. If the origin of this single kind of cell involved hundreds of genes being turned on at the same time, then how could pregnancy happen? It would require hundreds of mutations arising simultaneously across the entire genome.

For Lynch to answer his questions would require looking at each of those hundreds of genes that make decidual stromal cells.

To consider Lynch's next step, we need to pause and consider what would make genes turn on in order to transform a cell into a decidual stromal cell. Recall that there are molecular switches across the genome that, under the right circumstances, turn genes on and off. Most of these switches lie right next to the genes that they activate. Since progesterone is the trigger for the formation of decidual stromal cells, then we could reasonably assume that the switches would be responsive to it. The genetic switches would be tethered to a sequence that recognized progesterone. When progesterone was present, the switch would activate and the gene would make protein.

This insight gave Lynch the clues he needed to probe the genome. He could look for the telltale signature of genetic switches that had, as part of their sequence, a region that recognized progesterone. This region would have a sequence that the hormone could bind to, so with any luck he could find them in a comparison of his genes within computer databases.

And that was exactly what he found. Almost all of the hundreds of genes that make decidual stromal cells had a switch that responded to progesterone. This discovery, while interesting, did little to answer the question that got Lynch into all this in the first place. Somehow, during the origin of pregnancy, hundreds of genes had to become active in response to progesterone. Since hundreds of genes are turned on in response to progesterone, hundreds of switches that respond to progesterone had to exist across the genome, near each of the genes that is activated by the hormone. This was no simple mutation of DNA, like changing a single letter in the code. Lynch was seeing a batch of letters that had to change simultaneously in hundreds of places across a genome to make decidual stromal cells. The implausible just got impossible.

As each new experiment made the origin of the cells ever more unlikely, Lynch returned to the structure of the genetic switches themselves. Perhaps something they all shared would offer an explanation? Looking in detail at the sequences, he used a computer algorithm to see if there was any shared pattern. A simple gene sequence emerged, one that was shared by virtually all the switches. Running the sequence across a huge database of all known sequences, he found the answer: each genetic switch had the telltale signature of a jumping gene, the type of gene that McClintock found first in corn. These genes, as we saw earlier, make copies of themselves to insert all over the genome. McClintock had seen them as great disrupters—that is, when they hop and insert into another gene, they can disrupt the function of that gene and make a pathology. Lynch saw something else.

This simple linkage made possible a complex, seemingly impossible invention. Hundreds of genes did not have to mutate independently. Lynch saw that a mutation happened in a single jumping gene, turning a regular sequence into a switch that responded to progesterone. Then the mutation spread across the genome as the jumping gene with the switch duplicated, jumped, and landed in new places. Jumping genes distributed switches all over the genome very rapidly. When they landed next to a gene, that gene could now be turned on in response to progesterone. In this way, hundreds of genes gained the ability to be turned on during pregnancy. A genetic change, involving the coordination of hundreds of genes, could occur not by hundreds of independent mutations but by jumping genes carrying a single mutation throughout the genome. In this way, genetic changes could spread very quickly as genes jump, make copies of themselves, and land in different places.

Jumping genes are the ultimate selfish elements—they can duplicate and jump to spread and multiply across the genome. Lynch was finding that, on occasion, jumping genes can carry useful mutations that do dramatically new things.

There is a war going on inside the genome, between jumping genes and the rest of our DNA. That tension between a selfish gene and the forces that strive to control it occurs in genomes every day. It turns out that DNA has hidden mechanisms to control jumping genes. One of them involves a small DNA sequence that functions like a hunter-killer, able to silence jumping genes by attaching to the part of the gene that makes it jump, then literally bundling it up in protein so it cannot jump around. Neutered in this way, the gene doesn't jump; it stays put. This silencing mechanism can control jumping genes and stop them from taking control to the point of disrupting the workings of the genome. It may also serve to domesticate jumping genes. If a jumping gene contains a potentially useful sequence, the hunter-killer DNA can neuter the jumping ability and make it stay put to play a new role. It can silence the jumping part but keep the helpful mutation.

That is what Lynch found with his switches: each of the switches that made decidual stromal cells had a special sequence that looked for all the world like it originally came from a jumping gene. But the gene had one difference: a small stretch of DNA was missing, and not just any DNA—the DNA that caused the gene to jump. It was as if the code had been hacked to stop the gene from jumping and keep it in place to do its work of making decidual stromal cells. With its springs clipped, the no-longer-jumping gene was put to work where it landed.

What Lynch saw in pregnancy is a window into a much larger world. Genomes are at war with themselves: between jumping genes and the forces that try to contain them. Out of this struggle comes invention, where a single mutation can spread across the genome and, over time, bring about a revolution.

These shifts are a far cry from Goldschmidt's hopeful monsters. A revolutionary mutation doesn't have to arise in a single step. An incremental change can arise in one place in the genome and, if tethered to a jumping gene, spread and be amplified over time in subsequent generations.

But the war inside the genome extends even wider. And pregnancy, again, reveals how.

Hacking the Hackers

In the placenta, right at the boundary between the fetus and the mother, one protein has a very special role to play. Syncytin sits at this interface and serves as a molecular traffic cop as the mother and fetus exchange nutrients and waste products. A number of observations show that this protein is vital for the health of the embryo. When a group of scientists made a mouse with a defective syncytin gene, the mice grew and lived normally, but they couldn't reproduce. After fertilization, the placenta would fail to form, and the embryo would not survive. Lacking syncytin, the mother could not make a functional placenta, and the embryo had no way of obtaining nutrients. Defects in syncytin also cause a wide range of problems in pregnancy in people. Women with preeclampsia have a defective syncytin gene; they make the protein, but it cannot do its job well. This sets off a chain reaction in the placenta that leads to dangerously high blood pressure.

A biochemistry laboratory in France began to look at the structure of the protein by exploring the sequence of DNA that makes it. As we saw with Lynch's work, once a gene is sequenced, the code can be run on a computer and compared to databases containing other genes in living creatures. These pattern-recognition packages cross-check the entire gene as well as small stretches of it for any similarities to other genes that have been sequenced. Over the past few decades, databases have been filled with millions of sequences of proteins and genes for everything from microbes to elephants. These comparisons have revealed that many genes are part of the duplicated gene families that we saw in Chapter 5. In the case of syncytin, the researchers were looking for similarities to other proteins that might give clues to how syncytin works during pregnancy.

The searches were revealing a puzzle. The database hunt showed that syncytin had no similarities to proteins in any other animals. It didn't look like anything in plants or bacteria either. The computer match was as bewildering as it was surprising: the sequence of syncytin looked for all the world like a virus and was identical in places to HIV, the virus that causes AIDS. Why would a virus like this have any similarity to a protein in mammals, let alone one that is an essential part of pregnancy?

Before exploring syncytin, the researchers needed to become experts on viruses. Viruses are devious molecular parasites. They have genomes stripped of everything but the machinery needed for infection and reproduction. They invade cells, enter the nucleus, and enter the genome itself. Once in the DNA, they take over and use the host's genome to make copies of themselves and produce viral proteins instead of those of the host. With this infection, a single host cell becomes a factory to make millions of viruses. For a virus like HIV to spread from one cell to the next, it makes a protein that causes the host's cells to stick together. The protein brings the cells together and makes pathways for the virus to move from cell to cell. To do this, the protein sits at the interface between the cells and controls the traffic between them. Does this sound familiar? It should, because syncytin does the same thing in the human placenta. Syncytin brings cells together in the placenta and controls the traffic of molecules between the fetal and the maternal cells.

The more they looked, the more the team found that syncytin is essentially a viral protein that has lost its ability to infect other cells. This similarity between a mammalian protein and a virus led to a new idea. At some point in the distant past, a virus invaded our ancestors' genome. That virus contained a version of syncytin. Instead of commandeering our ancestors' genome to make endless copies of itself, the virus became neutered, lost its ability to infect, and then was put to work by a new master. Our genome is in a continual war with viruses. In this case, by mechanisms we have yet to understand, the infectious part of the virus was knocked out, and the virus was put to use making syncytin for the placenta. Viruses brought the protein to the genome, and the attacker's genome was hacked to be useful for the host.

The scientists then looked at the structure of syncytin in different mammals and found that the version in mice is different from that in primates. Comparing the databases, they saw that different viral invasions are responsible for the syncytins in different mammals. The primate version arose when a virus entered the common ancestor of all living primates. The syncytin of rodents and other mammals came about from a different event, leading to their versions of syncytin. The end result is that primates, rodents, and other mammals have different syncytins derived from different invaders.

Our DNA is not entirely an inheritance from ancestors. Viral invaders have inserted themselves and been put to work: our ancestors' battles with them have been one of the many roots of invention.

Zombie Memories

When Jason Shepherd was a child growing up in New Zealand and South Africa, he so pestered his mother with questions that she finally told him he needed to become a scientist to find his own answers. By the time he graduated from high school, he had decided to enter medicine. He began a crash program to give him both premed and medical training in a short few years. In the first year of the program, he encountered Oliver Sacks's classic *The Man Who Mistook His Wife for a Hat*. That single book changed his life. Inspired by Sacks, he left the program and launched a new career studying the molecules and cells that make our brain work. His quest, as he describes it, became to find out what makes us human. Memory, and its loss, became Shepherd's scientific quarry. Our ability to recollect the past defines much of how we learn, relate to others, and function in the world. This is no esoteric subject. One of the great challenges we face as a society is neurodegenerative disease. As we lengthen our lifespan, the aging brain serves as an ever more critical barrier. Loss of memory and cognitive function are scourges with emotional, social, and financial tolls that are incalculably large.

In Shepherd's senior year in college, as he was looking for a paper topic for a course in neurobiology, he ran across an article on a gene called *Arc* that appeared to be involved in making memories. In mice, *Arc* is turned on as creatures learn. Moreover, it is active in the brain in the spaces between the different nerve cells. *Arc* seemed to fit the bill of a gene important in memory.

A few years after Shepherd's college assignment, technology had evolved to the point where researchers were able to make mice lacking the *Arc* gene. The mice survived but had a number of defects. When offered a maze with cheese in the center, they could solve the maze, but they could not remember its structure the next day. This is something that mice with normal memories can often do. In test after test, the mice revealed a specific deficit in forming memories. Mutations of *Arc* in humans are known to be associated with a range of neurodegenerative disorders, from Alzheimer's to schizophrenia.

Memory and the *Arc* gene became the focus of Shepherd's career. He went to graduate school to study *Arc* with one of the biologists who had first explored its role in behavior. Then, after graduating, he did his postdoc training with the scientist who discovered where the *Arc* gene lies in the genome. Shepherd had *Arc* on the brain both literally and figuratively.

Building his own lab as an independent scientist at the University of Utah, Shepherd devised experiments to understand how the protein of *Arc* works. Clearly it is involved in conveying signals from one nerve cell to the next, and that signal is important in memory and learning. He would find answers to his questions by purifying the protein and then analyzing its structure.

Purifying a protein involves a number of steps to strip away everything in a cell but the protein of interest. The process begins with chemically macerating the tissue—in this case, brains—into fluids, then treating them successively to isolate the desired protein from all the others present. The protein soup is run through a series of tubes with each pulling out different contaminants. In one of the final steps, the fluid is run through a glass column packed with a special gel. The gel removes the final contaminants and other proteins, and the fluid that makes it through contains only the purified protein. Shepherd went through each step, getting small amounts of liquid to process along the way. He poured the fluid into the last glass column and—nothing. Nothing came out of the column. He changed the gel to a fresh batch. Again nothing came through. Clearly something was clogging it. The team tried new columns, but the tubes were still clogged. They tinkered with concentrations of different fluids. The clogs remained.

Shepherd's lab technician had a hunch. Maybe there was something special about the *Arc* protein that clogged the columns. Instead of being an artifact, perhaps this was saying something about the structure of the *Arc* molecule itself. Shepherd and his assistant took the clogged fluids to an electron microscope, where they could see the structure of the proteins on a computer screen at ultrahigh magnification. The structure was so surprising that Shepherd exclaimed, upon seeing it, "What the hell is going on?"

Arc was forming hollow spheres, and these spheres were so big that they got stuck in the spaces inside the gel filter. He had seen versions of these spheres before, in his premedical training. The structure of the spheres was identical to those made by some viruses as they move from cell to cell to infect them.

Shepherd works in the research wing of the University of Utah Medical Center, so he went across the building to visit a team that studies the virus that causes HIV. HIV moves from cell to cell by forming a protein capsule that conveys its genetic information. Showing the microscopic images to the virology team, Shepherd left it to the scientists to figure out what the curious spheres were. The HIV researchers thought they were from a virus like HIV. They couldn't find any difference between the *Arc* capsule and those made by the HIV virus. Both were made of four different chains of proteins, and both had the same molecular structure, even down to the atomic architecture of the bends and folds. Much like anatomists studying and naming bones, biochemists have their names for structures as well. A bend in the molecular structure known as zinc knuckle is one characteristic of HIV. *Arc* had that too.

It became clear that the *Arc* protein was virtually identical to viruses like HIV. And both molecules functioned in the exact same way—they conveyed small bits of genetic material from one cell to the next. Syncytin, as we have seen, is also HIV-like, albeit in different ways.

Working with geneticists, Shepherd's team mapped the structure of *Arc* DNA and scoured the genome databases of the animal kingdom for other creatures that have it. In tracing the structure and distribution of the gene, a story of ancient infections emerges. All land-living animals have the *Arc* gene; fish do not. This means that about 375 million years ago a virus entered the genome of the common ancestor of all land-living animals. I like to think that it was a close relative of *Tiktaalik* that got the first infection. Once the virus joined the host, it carried with it the ability to make a special protein, a version of *Arc*. Normally the protein would be used to allow the virus to move from cell to cell and spread. But in this case, because of where it entered the fish's genome, it made that protein become active in brains and enhanced memories. The individuals with the virus were the recipients of a biological gift. The virus was hacked, neutered, and domesticated for a new function in brains. Our ability to read, write, and remember the moments of our lives is due to an ancient viral infection that happened when fish took their first steps on land.

Excited to present his results, Shepherd went to a conference on neuroscience and behavior. Before he spoke, he heard a scientist who works on fruit flies give her talk. She showed that the flies have *Arc*. Fly *Arc*, like ours, is active in spaces between neurons. Moreover, fly *Arc* forms hollow capsules that convey molecules from one nerve cell to the next. But fly *Arc* looks like a different virus from the one in land-living animals. Theirs came from a separate encounter with viruses.

How does a genome domesticate a virus and put it to work rather than allow it to infect? The answer is not clear, but there are many different ways this might happen. Think of the fate of both a virus and a host under a few different circumstances. If the virus is very infectious, the host will die and the virus will not pass from generation to generation. If the virus is relatively benign, or beneficial, it will enter the genome and reside there. If it makes it to the genome of a sperm or egg, the virus will pass its genome on to offspring. Over time, if the virus has a very beneficial effect, say by making creatures with more efficient placentas or better memories, natural selection can hone it to stay put and do its job ever more efficiently.

The genome is the stuff of B movies, like a graveyard filled with ghosts. Bits and pieces of ancient viral fragments lie everywhere—by some estimates, 8 percent of our genome is composed of dead viruses, more than a hundred thousand of them at last count. Some of these fossil viruses have kept a function, to make proteins useful in pregnancy, memory, and countless other activities discovered in the past five years. Others sit like corpses where they attached to the genome only to be extinguished and decay.

A struggle is going on inside genomes. Some bits of genetic material exist to make ever more copies of themselves. They can be foreign invaders, such as viruses that enter the genome to commandeer it. They can also be innate parts of our genome, such as jumping genes that proliferate and insert everywhere. Occasionally, when these selfish genetic elements land in a special place, they can be put to use to make new tissues, such as the endometrium, or to allow for new functions, such as memory and cognition. Genetic mutations can spread far and wide across the genome in a small number of generations. And if viruses occupy different species, similar genetic changes can arise in different kinds of creatures independently.

My Thursday teas with Mayr continued for another two years after the Goldschmidt faux pas. During those later meetings, I discovered that Mayr had grudging respect for Goldschmidt's attempt to unite experiments in genetics and developmental biology with the major events in the fossil record. By the mid-1980s, he knew that a revolution was coming from molecular biology and so encouraged the graduate students in his orbit to keep current in that area of research.

As Lillian Hellman might have said in this context, nothing ever begins when, *or where*, you think it did. Genomes are not static strands; they are ever twisting and turning while viruses attack and other genes jump. Genetic mutations can spread across the genome and among different species. Changes to the genome can be rapid, similar genetic changes can happen independently in different creatures, and the genomes of different species can blend and merge to forge biological inventions.

7

Loaded Dice

I PAID THE BILLS during my last year of graduate school by working the graveyard shift as a security guard in the chemistry department while serving as a teaching assistant during the day. Since the chemistry buildings at three a.m. were populated by only a small number of night owls, I'd make my rounds and then enjoy quiet nights delving into the classic literature in paleontology. At the end of my shift, I'd do my own research and then assist teaching paleontology in a large lecture class. This time gave me exposure to great ideas and debates. It didn't hurt that my main teaching gig was serving as one of the flock of assistants to the late Stephen Jay Gould in his popular history of life class.

By the mid-1980s, Gould had emerged as a major public figure, using his background as a paleontologist to dive into controversies with radical stances on the ways new species emerge and how evolutionary change comes about. His college class was composed of around six hundred students who, taking it as a distributional requirement, were unlikely to become science majors. This audience proved an ideal focal group for Gould to try out his new theories and presentations. Every Tuesday and

Thursday in the fall he held forth, lecturing with dramatic flourish to undergraduates who either sat rapt in the front rows or sprawled sleeping in the rear ones.

At the time, Gould was thinking about cataclysms that happened in the history of life. Five times in the past 500 million years, long-dominant species from around the globe had abruptly disappeared. The most famed of these extinctions is the one that caused the demise of the dinosaurs. About 65 million years ago dinosaurs, marine reptiles, pterosaurs, and many kinds of invertebrates that lived in the oceans were extinguished. Plant diversity declined worldwide as well. Evidence in the rocks revealed the likely cause—a large asteroid hit the Earth, changed global climate dramatically, and led to the collapse of ecosystems worldwide, with numerous animals going extinct rapidly. Removing dinosaurs and other creatures paved the way for mammals to expand into a world depleted of large predators and competitors.

In one lecture, Gould posed "what if" counterfactual questions. What if an asteroid had not crashed into Earth and dinosaurs and other creatures had survived? What if many of the seemingly contingent events of history hadn't happened—what would the world look like? The lecture was before winter break, and after an annual viewing of Frank Capra's *It's a Wonderful Life*, Gould drew an analogy from the film. The film's hero, George Bailey, is ready to jump off a bridge to end his life when an angel intervenes, giving him a chance to travel in time to see how his suicide would affect his hometown. Without Bailey, Bedford Falls, New York, is altered for the worse. Gould substituted an asteroid impact for George Bailey, and life on Earth for the residents of Bedford Falls. If an asteroid had not hit the Earth 65 million years ago, the dinosaurs would likely have persisted, and mammals might never have flourished. In fact, we might not even be here were it not for that random collision of a rock with our planet.

That collision is just one of a string of innumerable, seemingly contingent events that happened over the past four billion years in order for us to be here today. Just as our own personal lives have been shaped by numerous random encounters, conversations, and opportunities, so has the history of life been shaped by changes to the cosmos, planet, and genomes. Gould's lecture would later become fodder for his best-selling book *Wonderful Life*. In it, Gould generalized this "what-if" thinking to great moments in the history of life. The natural world we see around us today, including our own existence, is the product of eons of contingent events. Replay the tape of life with any one of them different in even a small way, and the world—including our very presence in it—would be drastically different.

Recent science, coupled with nearly a century of work, points to a different conclusion altogether. Replay the tape of life with different contingent events, and perhaps some outcomes wouldn't be so different after all.

Degenerates

Sir Ray Lankester (1847–1929) was a giant of a man in both height and circumference. He was garrulous, highly opinionated, and combative. Raised by a physician who encouraged him to explore the natural world, Lankester was primed from his childhood for a career as a scientist, ultimately training at Oxford in the 1860s with some of the leading lights of the day.

After Darwin published *On the Origin of Species*, Thomas Huxley defended Darwin so vociferously that he became known as

Sir Ray Lankester

"Darwin's bulldog." Not surprisingly, Lankester found his way to Huxley. Lankester was so pugnacious that recent historians of science have given him the moniker "Huxley's bulldog." He had such a proclivity to argue, often angrily, that even Huxley himself had to calm him down on occasion.

Lankester became a debunker of claims of the paranormal, which were rampant in the Victorian times in which he lived. He famously exposed the American medium Henry Slade during a séance in London. Slade was known for pulling a slate and chalk out from under a table during a séance to reveal messages from the spirit world. Using his size as a weapon at one such séance, Lankester grabbed the slate before a show to reveal prewritten messages. Lankester was zealous enough to pursue a criminal prosecution against Slade.

The same boisterous commitment to skepticism that exposed hoaxes propelled Lankester's science. After Oxford, he trained as an anatomist at the Stazione Zoologica in Naples and became an expert in marine clams, snails, and shrimp. In his hands, the anatomy of these creatures held surprises, and he was comfortable pursuing the trail of evidence no matter where it led.

After Darwin, anatomists looked for similarities among species that could be clues to their ancestry. Recall that Darwin's reasoning was that anatomical similarities among species are evidence that they share a common ancestor. Huxley identified certain groups of fish that were close relatives of limbed animals because their fins had versions of arm bones inside. Likewise, he and others used anatomical similarities to show that birds and mammals have affinities with various reptiles. This reasoning made specific predictions: forms that are closely related should have more similarities than ones that are more distantly related.

Lankester saw something else: he focused on an observation that was either unseen or ignored by other scientists. In his work on marine animals, he found that many species evolved not by gaining new traits but by losing them. Shedding structures and becoming simpler, or "degenerating," as Lankester called it, opened up new ways of living. He noticed that when creatures evolve a parasitical lifestyle, they become simpler and lose body parts, often entire organs. Shrimp are creatures with tails, shells, eyes, and nerve cords, but parasitic shrimp that live in the guts of other creatures are almost unrecognizable as such. They shed the shell, eyes, and even many of their digestive organs.

Lankester's study of degeneration led to an even deeper and more important observation. Parasitic shrimp, no matter where they live on the planet or what part of their host they are specialized for, whether in fish guts or gills, always lose the same body parts. The same is true for many other degenerates.

Cave-dwelling animals, be they fish, amphibians, or shrimp, lose organs to become more efficient at living in dark caves, presumably saving the energy that would be spent building and maintaining useless organs. Surprisingly, different species evolve in the same way independently: they become colorless and lose their eyes and often reduce the size of their appendages.

Perhaps one of the most obvious cases of degeneration is snakes that lose limbs, except for a small nubbin seen in some species. The snake body plan doesn't involve only loss; bodies also get longer by the addition of vertebrae and ribs. This is part of the snake lifestyle of locomotion by slithering. Limbs would simply be in the way in this kind of movement.

A snakelike body, as Lankester knew, is not limited to snakes. A number of different lizard species have highly reduced limbs and long bodies. One distantly related group of reptiles, amphisbaenians, have long bodies and no limbs. You would be excused for mistaking them for snakes or lizards, but their head anatomy is very different. Even amphibians get into the game. Amphibians known as caecilians have long bodies and no limbs. Here is the same trait, and the same way of evolving, arising in different animals many times.

Independent inventions are a common pattern in the world of human innovation as well. Whether it is the telephone, the yo-yo, or the theory of evolution, ideas and technologies have a habit of arising with different inventors around the same time. Perhaps an idea is in the air because the timing is right, is an obvious improvement on an existing technology, or is caused by some deep regularity in the way that invention happens. Whatever the cause, "multiples" are so widespread as to be the rule in some fields of human endeavor. The same is true for parts of the living world.

Biological multiples can reveal the inner workings of nature.

To see how, we need to return to Auguste Duméril's humble little animals.

A Salamander View of the World

With his soft-spoken and collegial manner, nobody could ever confuse David Wake, of the University of California at Berkeley, with Ray Lankester. The impact of Wake's work since the 1960s has been just as profound, however. While Lankester's métier was marine animals, Wake has dedicated his scientific life to understanding salamanders.

We should be so lucky as to have some salamander biology inside us. Cut one of their limbs off and they can regenerate it entirely, including all muscles, bones, nerves, and blood vessels. Salamanders regrow damaged hearts, even spinal cords. They

David Wake looking for salamanders in Mexico

have remarkable inventions, ranging from different kinds of poison glands to the ways they capture food. For over four decades, students and senior scientists have traveled to Berkeley from around the world, from dozens of different countries, to learn the biology of salamanders. Wake is a modern-day Duméril, finding surprising biological insights in simple-looking salamanders.

As we have known since Duméril, salamanders are typically born in one environment, then, as they grow, they switch to a new one. Many species hatch in water and then metamorphose to live on land. The transition to land involves wholesale changes in how the animals live, especially how they feed.

Generally speaking, there are two kinds of predators. Most bring their mouth to the prey: lions, cheetahs, and crocodiles snap or bite as they chase prey or silently wait for them to pass by. Other predators acquire their food in the opposite manner, by bringing the prey to their mouths. Adult salamanders belong to this category.

While in water, salamanders bring insects and tiny arthropods into their mouth by sucking them in. Tiny bones at the base of their throat, as well as others at the top of the skull, expand the mouth cavity and create a vacuum that pulls water and prey inside. While this strategy works well for amphibians in water, it is a nonstarter on land. Land animals would require a jet-engine-strength vacuum larger than their entire bodies to create enough suction to pull heavy prey through the air and into their mouths.

Salamanders employ many tricks to get prey inside their mouths on land. Some species project their tongue outside the body, capture insects, and reel them in. They flip their tongue almost half the length of their body, shooting a sticky pad to catch small insects and convey them to the mouth. Two kinds of features allow salamanders to accomplish this feat: mechanisms that project the tongue and those that retract it. This specialized tongue is one of nature's most remarkable inventions, and while it may seem painfully esoteric, it holds general surprises for understanding life on Earth. Since the beauty and importance of this system emerges from the anatomical details, we need to dig into some salamander anatomy.

To start thinking about salamander tongue flipping, try sticking your own tongue out. A complex interplay of muscles makes the motion possible. Our tongue is essentially a set of muscles wrapped with connective tissue and covered with taste buds. A series of other muscles connect the tongue to the bones of the jaw and throat. Sticking your tongue out moves muscles internal to the tongue—those that change their shape from soft to rigid and from flat to elongated—as well as exterior muscles that attach to the tongue, to pull it outside the mouth. One of the main muscles that pulls the tongue outside the mouth attaches to the base of the chin and connects to the base of the tongue. When this muscle, the genioglossus, contracts, your tongue sticks out.

Humans use the genioglossus muscle to talk and eat. In fact, a modification of the genioglossus is sometimes used as a surgical remedy for snoring. Tightening the muscle moves the tongue's resting position forward, away from the throat. This adjustment stops the tongue from obstructing the airway during sleep, thus preventing snoring and also, it is hoped, sleep apnea.

While we humans are justifiably proud of our ability to talk, of which movements of the tongue and the genioglossus muscle are such vital parts, we would be hopeless trying to capture flying insects. Tongues like ours protrude neither far enough nor fast enough to capture anything. That's likely a good thing, given our social norms and food choices, but this state of affairs won't work for salamanders.

Many salamanders have a genioglossus muscle as well, and it plays a role in feeding. A number of species modify the genioglossus into a long strap that, when contracted, enables the tongue to protrude outside the mouth. This kind of tongue projecting is the most common among salamander species. In the Olympics of tongue projection, however, it wouldn't even get to a preliminary heat: it is great, but nowhere near as incredible as other mechanisms. The speed at which the genioglossus muscle can contract sets a physical limit for how fast this system can work. While it is fast, it is not fast enough to capture many rapidly flying insects.

Members of the salamander genus *Bolitoglossa*, one of Wake's specialties, can protrude their tongue half a body length and then retract it in less than two-thousandths of a second. Watching them feed is a bewildering experience. The tongue moves so fast that the motion can barely be perceived, even in slow-motion videos on YouTube. What is mind-boggling is that no salamander muscle can contract as fast as their tongue projects; they are shooting their tongues faster than the speed limit of the muscles themselves. These salamanders seem to break the laws of physics.

David Wake and one of his graduate students in the 1960s, Eric Lombard, focused on these tongues in a nearly ten-year effort to understand how they work and, importantly, how they came about. They dissected tongues from different species and looked carefully at every muscle, bone, and ligament. They manipulated different bones and muscles with tweezers to see if they could simulate the motions. Decades later, one of Wake's students filmed high-speed movements of the tongues to see how the muscles and bones worked together to seemingly do the impossible.

Wake discovered that salamander tongues function like an

extremely intricate biological gun. Highly specialized salamanders don't just stick their tongue out. Their tongue shoots out of the mouth like a bullet tethered to a string. If that is not strange enough, the projectile that the salamander shoots is the small bones of its gill apparatus that lie attached to a sticky pad. They literally propel parts of their gills up to half a body length in the blink of an eye. Then, just as remarkably, the tongue snaps back into the mouth just as fast as it was ejected.

In salamanders with projectile tongues, the genioglossus is completely lost. That muscle contracts too slowly and would just get in the way as the projectile shot out. Also, in most salamander species, gill bones lie fixed on either the side of the head to serve as a base for the gill filaments. Salamanders with projectile tongues do things differently. The gill bones are freed from the skull and are attached to the tongue to become the projectile that gets shot like a bullet.

To conjure an image of salamander tongue projection, imagine shooting a watermelon seed by squeezing it between your thumb and forefinger. The seed is slippery and tapered. When you squeeze your fingertips on it, the seed shoots out quickly and far. The same is true for salamander tongues. Elaborate muscles serve as the squeezers, and the bony rods of the gill apparatus become the lubricated and tapered surfaces. When the muscles contract, off the bones go, much like the watermelon seed.

In projectile tongues, two gill bones are expanded to look like a tuning fork with the tines facing the tail end. These long rods are tapered and lubricated, much like the watermelon seed. Wrapped around these rods are constrictor muscles that run along their length. When these muscles fire, they squeeze the rods and shoot them out of the mouth. The end result is that the tongue pad and the gill bones shoot to their target. If the process works, the insect is captured by the pad and returned to the mouth.

It would do a salamander no good to shoot its tongue and catch an insect but not be able to return either the prey or the tongue to its mouth. While the thought of a salamander unable to reel in a jangle of tongue might be comical, this state of affairs would be deadly. Exposed to predators and unable to acquire more food, the animal would almost certainly die. The solution is clever. In all salamanders, the abdomen is swathed in muscles that extend from the hip all the way to the gills. These muscles usually work to support the body. In the species with the most projectile of tongues, fibers of the two sets of muscles merge, making a single muscle that runs from the pelvis to the specialized gill bones. Imagine a giant spring: when the gill bones are shot out, the muscular strap stretches to recoil the apparatus.

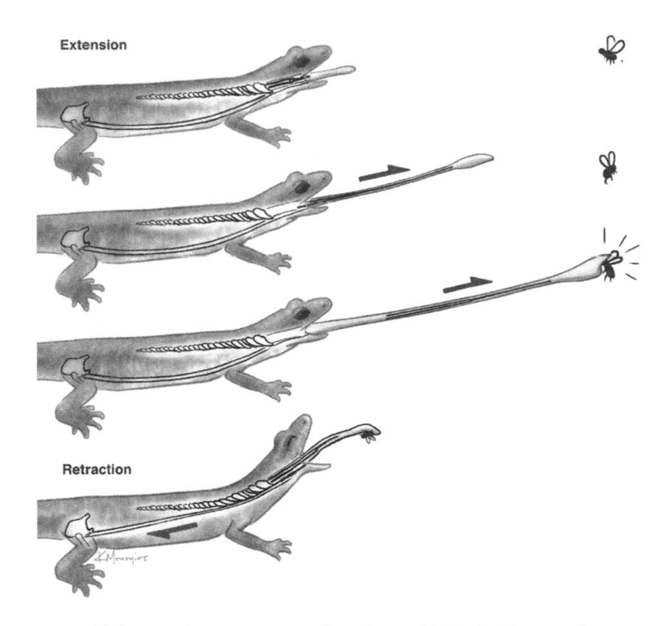

Salamander tongue projection, a biological marvel

The origin of this complex biological organ involved not the creation of new organs, or even bones, but the repurposing of ancient bones and muscles in novel ways. The muscles that propel the tongue are those that other salamanders use in swallowing. Bones that formerly supported the gills became tapered at one end to become the bullets. The genioglossus muscle has been lost to let the projectile fly far. Abdominal muscles have been fused to make the spring that retracts the tongue. This repurposing has made a natural wonder, a highly intricate invention involving many parts.

While the salamander tongue is a marvel unto itself, something even more extraordinary emerged from another area of Wake's research.

One of Wake's specialties is using DNA to decipher the salamander family tree—exploring how different species are related to one another. Following the tradition that began with Zuckerkandl and Pauling, he compares gene sequences among species so he can assess where and when they evolved. Using tissue samples taken from almost every salamander species, Wake composed the most definitive family tree for salamanders to date. Even he was shocked by the result.

The salamanders with the most extreme projectile tongues are not closely related to one another. In fact, these species were so far apart on the family tree that they lived hundreds of miles away and had different ancestors. The invention of a projectile tongue, an intricate biological novelty involving many coordinated changes across the head and body, came about at least three times independently, maybe even more. In all cases the genioglossus was lost, the gill bones were modified to be projectiles, and belly muscles were converted to a spring to return the projectile to the mouth. These tongues are examples of Sir Ray Lankester's multiples on steroids.

The independent invention of this highly specialized organ is no accident. All species that have this trait share several things. Most salamanders use the gill bones in breathing, to expand the mouth and pull air inside the lungs. And they use those gill bones extensively in larval stages to feed: movements of these bones generate the suction necessary to pull food inside. If gill bones are needed to breathe and feed, then how could they be used in tongue projection? Species with the most extreme tongue projection have neither lungs nor larval stages. Having lost both, the gill apparatus no longer has these competing functions and can serve in a new one, as a missile to catch prey.

But how do multiples emerge? What do they tell us about the inner workings of living things?

The Mess Is the Message

Scientists, like most humans, hate messes. Scientists love graphs where the points fit neatly on a line or a curve. We crave experiments that are definitive. Our ideal observations are neat, tidy, and uniformly follow a prediction. We love signal and loathe noise.

Studies of the tree of life are no different. Building the family tree of life is a bit like devising a key to identify species in the wild: we look for unique features that animals share. The more unique features a species has, the easier it is to differentiate that species from others. Everybody can tell the difference between gulls and owls, for example. Each has features that serve as identifiers, whether the round face of owls or the beak and body coloration of gulls. Consistency lies in having features, from anatomy to DNA, that are shared by different groups of creatures. People share features not seen in other primates, primates share features not seen in other mammals, mammals share features not seen in most reptiles, and so on.

Ray Lankester uncovered a chicken-and-egg problem: How do we distinguish similarities that evolved independently, or multiples, from similarities that reflect true genealogy? If salamander tongues, with all their intricate details, could come about independently, how can we ever have confidence that the presence of any trait provides evidence of relationship? The reality is that, in salamanders, tongues are just one part of the story. Multiples are seen in organ after organ.

So how does the world's leading expert on salamanders look at their evolution? David Wake, like most others in the field, has practically given up on using anatomy as an indicator of relationships. Why? No matter how many data are collected, it is very clear that salamanders in different parts of the world, at different times, came up with the same designs independently.

Perhaps the messiness in biological multiples is not a mere annoyance but a window into something fundamental. Maybe what we see as the noise is really the signal. What if certain ways of evolving are not contingent?

Multiples arise in living things in one of two ways. The first is the existence of a limited number of solutions to a problem. Take flying as an example. Any creature that flies needs a big surface area to produce lift, so flying creatures all have wings. The wings of birds, flying reptiles, bats, and flies look similar, but they have different structures inside and different histories that we can trace. The configuration of bones in the wing of a bird is different from that in a bat or a pterosaur. In a bat, the wing is a membrane that stretches between five elongated fingers, while in a pterosaur the wing is supported by a very long fourth digit. Insect wings are different still, being supported by completely different types of tissue. Physical necessity and history merge to produce these structures—each structure is a wing but is configured differently to reflect the different evolutionary histories of mammals, birds, reptiles, and insects.

Examples of these kinds of physical necessities abound; they were often called "rules" by early anatomists. Allen's Rule, formulated by Joel Asaph Allen in 1877, held that warm-blooded animals living in colder climates will have shorter appendages (limbs, ears, noses, and the like) than those living in warmer ones. The explanation is heat loss—animals that have elongated appendages would lose more heat than those that don't. Similarly, Bergmann's Rule, named after Carl Bergmann in 1844, referred to the observation that animals living in colder climates are on average larger than those in warmer ones. Heat loss is the constraint here, too, because small animals have a proportionally greater amount of surface area by which to lose heat. Both Allen's Rule and Bergmann's Rule generally hold true in different species living in different places.

There is another way multiples can occur. Darwin recognized that no two creatures in a population are alike, and that some kinds of variation can make an organism more successful in its environment, by having more offspring and being more robust. Those differences are the basis for evolution by natural selection: as long as you have variation in a population and some of it affects the success of the creatures in their environment, evolutionary change is an inevitable outcome. But natural selection can act only on the diversity that exists in a population. If there are no differences among individuals, there can be no evolution. And what if the variation is biased in some way? What if the genetic and developmental recipes that build bodies and organs can produce certain designs more easily than others, or others not at all? If this is true, then knowing how animals build organs during development could help you predict how they would vary in populations and, as a consequence, the likely ways they could evolve.

Cold Feet

After finishing graduate school at Harvard, I moved west to the University of California at Berkeley to study in some of its famed campus museums of zoology and paleontology. After a few weeks on the scene, David Wake's infectious enthusiasm for salamanders drew me in, and I started to devise projects I could do with his team. I was drawn to California as much for a change of climate as I was for the museums and salamanders. Five years spent in Cambridge, Massachusetts, with summer fieldwork in Greenland and Canada, made me ready to get away from the dark and cold to bathe in some California sun.

That sunny bliss was not to be found. When I arrived, Berkeley was in one of the most severe cold snaps of recent memory. I was soon to learn that nothing, not even a tent in Greenland, is more frigid than California in the cold. Both houses and people, including me, lacked insulation. Pipes froze throughout the city, and water was rationed. Little could I have known it at the time, but that California flash freeze was to influence my own thinking about the history of life.

At some point during the freeze, I went into Wake's lab, if only to warm up and fill some water jugs. He was just off the phone with a colleague in the National Park Service at the Point Reyes National Seashore. The cold snap had hit the park's freshwater lakes hard, freezing them over for the first time in decades. The animals were as unprepared as humans for the drop in temperature. The purpose of the call was to inform him that thousands of salamanders had frozen to death in these ponds, and the park service wanted to know if we'd like to use them for the zoology museum's collection. The animals were already dead from a natural catastrophe, so why not see what science could extract from them?

We now had at our disposal over a thousand salamanders to study. At Harvard, I had studied salamander limbs, looking at how their hands and feet develop in embryonic stages. Given my interest, we developed a plan to look at the feet of these salamanders to assess the skeletons inside. With two feet per salamander, that came to roughly two thousand feet we could study.

My excitement over two thousand salamander feet was not an absurdity. I was right out of teaching in Gould's class and wanted to test the extent to which evolution is contingent or inevitable. We were seeing multiples everywhere, from tongues to degenerates, from salamanders to shrimp. In fact, the more people looked, the more they found. Wake discovered that salamander feet evolve in very specific ways, and as in the tongue system, different species evolve the same way independently.

By virtue of the freeze, we had thousands of feet from a single population of one species. Our idea was to look at their limb patterns to assess how they varied between individuals. This is the kind of variation that is the fuel for evolution by natural selection. We could now ask the central questions: Is the variation in populations biased in some way? Do multiples happen because the fuel for natural selection, variation among individuals, is not random? If all limb patterns are equally likely to happen, then we should see random variation in the huge sample size of the frozen salamanders from Point Reyes. But maybe some hidden internal bias to the variation nudges evolution in certain directions.

In over 200 million years of evolution, salamander limbs have evolved like Lankester's degenerates: they lose structures rather than gain them. Several features in their skeletons appear again and again, whether the species evolved in China, Central America, or North America. First, they tend to lose digits, and always the same ones. When salamanders lose fingers or toes, they always lose them on the pinky side, never the opposite one. The second pattern is that they tend to evolve by fusing the bones of their wrist and ankle. Salamanders normally have nine bones in their ankles. Specialized species tend to lose bones in a very specific way—they fuse adjacent bones. Where an ancestor once had two separate elements, a descendant may have one large one. What Wake noted was that these patterns of fusion appear to be nonrandom. Certain fusions happen again and again, while others never do.

In museums, zoos, or even in the wild, scientists almost never have access to one thousand skeletons of a single species. This number of specimens was a bonanza, because we now had the numbers to gather some real statistics and test ideas. We could see whether the variation was biased and so could influence how the salamanders evolved. The challenge was to see inside their feet.

We couldn't simply X-ray the limbs; their skeletons were made of soft cartilage that would be almost impossible to capture with standard medical X-rays. There were also too many individuals to put into a CT scanner; the cost would have been astronomical, and I didn't have salamanders on my health insurance plan. We settled on a technique whose results were as beautiful as the test was simple. We set up a series of baths of alcohol, water, and some chemical dyes. Over a period of a few weeks, we moved the salamanders from bath to bath, keeping them in each long enough to let the fluids diffuse into the tissue. The last bath contained a special blue dye that stuck to the cartilages, labeling all of them teal blue. Then, in the grand finale, we set

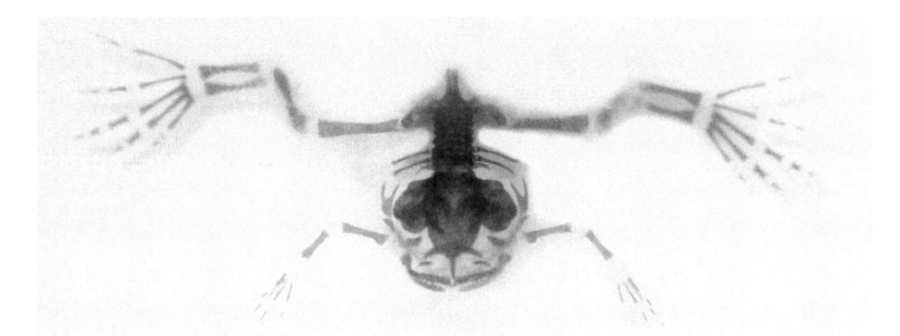

A frog whose body has been made clear and bones stained with dyes

the salamanders into a bath of simple glycerin, a clear viscous fluid. As the glycerin entered a specimen's body, it would make it as clear as glass. The process for a large salamander could take a few weeks. When we did it right, we ended up with something eerily beautiful. The animal was clear, and the skeleton was blue, as if it had been transformed into a blue skeleton in glass.

It took us two years to make one thousand of these preparations. We coded every limb in every specimen, recording every shape, fusion, and loss.

We found that the variation was not random: the answer was as clear as their bodies had become in the glycerin. Bones fused and specific digits were lost. What was more, we saw the same patterns of variation in this population of salamanders from Point Reyes that had been seen in species from China, Mexico, and even North Carolina. Some patterns of fusion were likely, others not. And in each case, we were seeing the same handful of patterns over and over again.

What can this tell us about salamander biology, let alone the contingency-inevitability dichotomy?

I had previously spent my graduate career studying how salamander limbs form during development. In looking at the making of their bones, there was a clear sequence to how the bones formed. The digits formed in a very precise order. Digit two formed first, followed by one, three, four, and five. I had seen this sequence before—it was exactly the order by which digits are lost in evolution. The first digit to be lost was the last to form; the next one, the second to last. It seemed that there was an organization to how digits were lost—last formed, first lost.

The cartilages in the wrists and ankles also develop in a well-defined sequence. They would bud from one another. One would form, then the next would bud from it. These two would separate as other newer elements budded. This budding and separation led to a complete pattern of nine independent bones. I had seen this before too. The elements that fused in the salamanders from different species were always ones that normally budded from one another.

Beneath this esoteric anatomy and development lies a simple and powerful notion. If you know how a salamander limb develops, you can predict how it is likely to evolve. The sequence by which the digits form, and the pattern by which the wrist and ankle bones bud from one another, determine that some pathways of change will be more likely than others. Last formed and first lost explains the variation we see in salamander digits. The fusions are not random, either. The elements that fused are ones that normally budded from one another in development.

Think about embryological development as a construction process. If you are a builder, the way you build a house, and the materials you use to construct it, can influence the kind of house you build. Some kinds of houses are more likely to be built than others. As we have seen with the frozen salamander feet, the

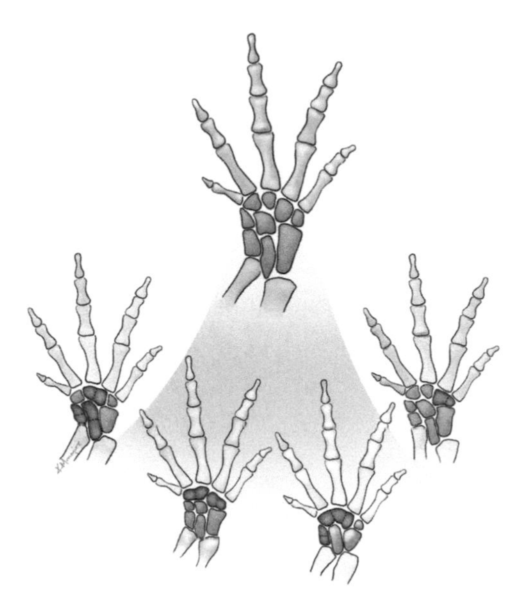

Salamander limbs evolve by losing elements. Shown here are the ways they fuse neighboring bones during evolution.

same is true with animals. The ways they are built make certain inventions and changes more likely than others.

For a long time, multiples, such as the bones in salamander feet, were seen as confounding artifacts in the history of life, almost like oddball one-offs. The more we look, however, the more we see that they are a regular part of the way invention happens. In many cases, they reflect deep rules of change, intrinsic biases that come from how species are built during development. If virtually every animal uses versions of the same genes—even whole genetic recipes—to build their bodies, then the existence of multiple after multiple in the animal kingdom should be no surprise. The arrival of great inventions in the history of life should be anything but contingent.

The path of evolution is not a continual line of progress fueled by random change. Over the course of history, different species often take different routes to the same place. To put this phenomenon in Gould's terms, replay the tape of life with different contingent circumstances, and important things would not be different, they'd be the same.

Ernst Mayr shared his own perspective during one of our teas. Riffing on Voltaire, he said that the results of evolution are not the "best possible world." Instead, they are the "best of the possible worlds." Genetics, development, and history help to define the kinds of changes that are possible.

Nature's Experiments

Nature does experiments for us. In fact, in some of them we can see the tape of life being replayed, just as George Bailey did on the bridge in Bedford Falls.

Lizards inhabit virtually all the islands of the Caribbean, from Saint Martin to Jamaica. With their lush forests, open plains, and beaches, these islands offer a range of productive environments in which lizards can thrive. Generations of scientists have found them a natural laboratory in which to study evolution. Much like the Galápagos for Darwin, each Caribbean island offers a way to assess how different lizards adapt to different environments. Ernest Williams (1914–98) was one of the great herpetologists of his generation. Building on the work of others, he noticed that various Caribbean islands have similar lizards on them. Lizards in forests are specialized to live in different parts of a tree: some in the canopy, others on the trunk, and others near the ground, at the base of the trunk. Every lizard that lives in a tree canopy, no matter on which island, is big, has a large head and a saw-like crest on its back, and is deep green. Every lizard that lives on the trunk is midsize, with short limbs, a short tail, and a triangular head. Every lizard that lives between the trunk and the ground has a large head and long legs and is mostly brown.

Mentored by Williams, my colleague Jonathan Losos has made these lizards the focus of his career. Losos used DNA techniques to explore the relationships among the lizards on various islands. Looking at their anatomy, you might expect that the big-headed lizards living in the tree canopies would be most closely related to big-headed ones on other islands, as would the short-limbed lizards on the tree trunks and the long-limbed ones near the ground. That is not what Losos found. Rather, the lizards on each island are most closely related to others on their own island. Each island has a genetically distinct lizard population and has been colonized separately. Castaways once landed on each island, and their descendants adapted to the conditions of their new home independently. Think of each island as a separate evolutionary experiment, in which lizards adapt to life on the ground, on tree trunks, on branches, and in the canopy. If each island is a separate experiment, then evolution has produced the same result over and over again. If the tape of history were replayed on different islands, evolution would have happened in the same way on each one of them.

The same situation is true at a grander scale for mammals. Marsupials have been evolving in Australia in isolation from the rest of the world for over 100 million years, producing diverse species with many different body shapes. The result is most definitely not random. There is a marsupial flying squirrel, a marsupial mole, a marsupial ground cat, and even a marsupial groundhog. And those are just the ones that are alive today—marsupial lions, wolves, and even saber-toothed cats are now extinct. Marsupial evolution on the isolated continent has often followed paths similar to those of the mammals in the rest of the world.

These natural experiments reveal that the history of life is not wholly a crapshoot of contingent events. The dice are loaded by the ways genes and development build bodies, by the physical constraints of environments, and by history. In each generation, organisms have inherited recipes—written in their genes, cells, and embryos—to build organs and bodies. This inheritance speaks to the future, as it can make certain pathways of change more likely than others. Past, present, and future merge in the bodies and genes of all living things.

8

Mergers and Acquisitions

SOMETIMES THE WORLD is not yet ready for a new invention or idea. Leonardo da Vinci (1452–1519) designed flying machines, including gliders, in the sixteenth century. They weren't made because neither the materials nor the processes to build them existed at the time. The history of life works the same way. Fish with lungs and arms thrived in ancient waters well before they took their first breaths and steps on hard ground. Creatures could never have survived on land because plants and insects were not yet abundant enough for any large animal to persist. Timing is everything in invention, whether it is in evolution, human technology, or even the struggles of a young scientist in the 1960s.

Lynn Margulis (1938–2011) studied microbial life at the University of Chicago and at Berkeley. In one of her first research projects, she looked at the diversity of cells in living creatures and proposed a new theory for how they arose. She wrote it up and received rejections from, as she once described it, "fifteen or so journals." Undeterred, she eventually found a home for the paper in a relatively obscure journal on theoretical biology.

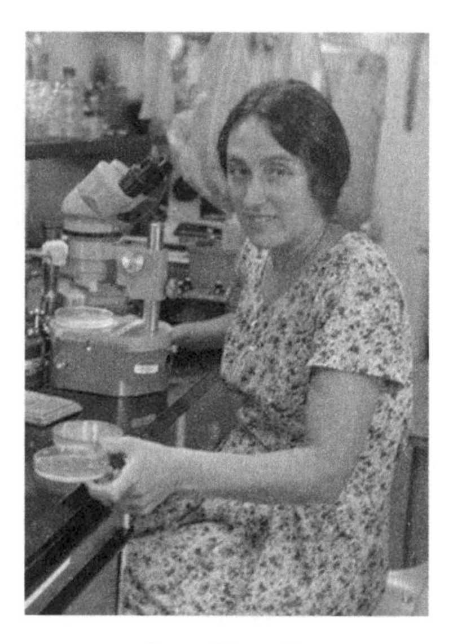

Lynn Margulis

Margulis's fearless persistence in the face of a chorus of negative reviews was breathtaking—here was a young female scientist at the start of her career set against an entrenched orthodoxy in a field dominated by males.

Margulis focused on the cells that make up the bodies of animals, plants, and fungi. These cells have a complexity to them that bacterial cells do not. Each one contains a nucleus, in which the genome resides. Surrounding the nucleus are a number of small organs, called organelles, that carry out different functions. The most prominent organelles are the ones that power the cell. Plants have chloroplasts that contain chlorophyll, which carries out the photosynthetic reactions required to convert sunlight to usable energy. Similarly, animal cells have mitochondria that generate energy from oxygen and sugars.

Margulis observed that these organelles look like mini-cells within the cell. Each has its own membrane around it, separating it from the rest of the cell. An organelle reproduces within the cell by splitting into two, or budding: first it becomes elongated and pinches in the middle like a dumbbell; then the two sides separate to form two new individuals. The organelle even has its own genome, separate from that of the cell nucleus. The genome of an organelle is very different from that of the nucleus, however. A DNA strand in the nucleus is coiled in on itself, but in mitochondria and chloroplasts, the ends of a strand of DNA close to form a simple ring.

Structured as they are with their own membranes, reproduction, and DNA organization, these organelles rang a bell for Margulis. She had seen these features before—in single-celled bacteria and blue-green algae. Bacteria and blue-green algae reproduce by budding, are surrounded by a similar membrane, and have a genome that looks much like that of chloroplasts and mitochondria. The organelles that power animal and plant cells looked for all the world to be more similar to bacteria and blue-green algae than to the nucleus of the cell in which they resided.

Using these observations, Margulis proposed a radical new theory of evolutionary history. Chloroplasts were originally free-living blue-green algae that got incorporated into another cell and were put to work as metabolic laborers to provide energy for it. Likewise, mitochondria were originally free-living bacteria that merged with another cell and were put to use powering it. Her radical notion was that in each case different individuals came together to make a new, more complex one.

Befitting a paper with fifteen rejections, Margulis's idea met with widespread scorn or complete indifference. Unknown to Margulis, sixty years earlier Russian and French biologists had independently proposed a similar notion that was ridiculed and remained hidden in obscure journals. But Margulis's fearless style, persistence, and creativity kept her idea alive as she spent several decades building more evidence and arguing tenaciously in public. Unfortunately, her efforts were to no avail. She remained on the margins of respectability because the similarities she was revealing did not convince the field.

Fortunately for Margulis, and for science in general, technology caught up with her idea. Once more rapid DNA sequencing methods were developed in the 1980s, the history of the genes inside organelles could be compared to those inside cell nuclei. The family tree that emerged was as beautiful as it was surprising. Neither mitochondria nor chloroplasts were genetically related to the DNA of their own cell's nucleus. Chloroplasts were more closely related to different species of blue-green algae than anything else inside the plant cell. Likewise, mitochondria were

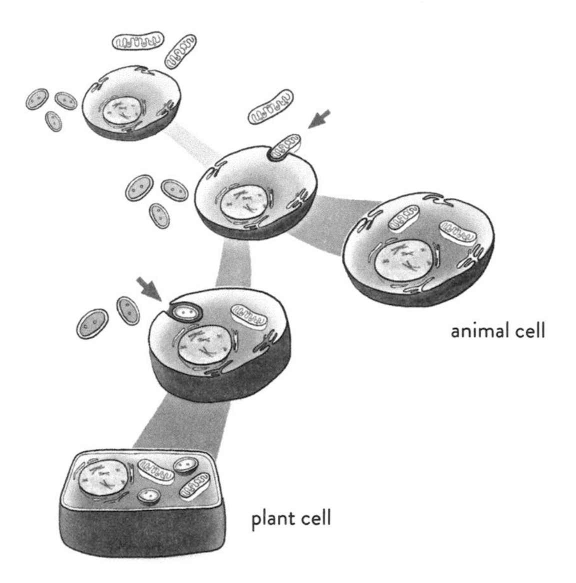

Evolution by combining: the origin of complex cells by the merger of two different kinds of microbes (arrows), one that gives rise to mitochondria (top), another to chloroplasts (bottom)

descendants of a species of oxygen-consuming bacteria and were unrelated to their nuclei. Every complex cell has two families of life inside it, one of its nucleus and another whose ancestors were once free-living blue-green algae or bacteria.

Recent DNA comparisons point to these kinds of combinations as being common events in the history of life. Cells unrelated to animals and plants, with different organelles, arose this way as well. For example, *Plasmodium falciparum*, the microbe that causes malaria, has a strange organelle that sits like a dunce cap on one side of the cell. It is used in a number of different metabolic processes. DNA sequencing shows that it was once a free-living algae. Because of its history as an individual cell, the organelle has distinctive molecules that lie on the membranes that surround it. Those molecules have been put to good use by medicine: they are a target that antimalarial drugs use on a search-and-destroy mission to kill malarial cells.

Margulis weathered her storm, but, sadly, her career ended in 2011, when she had a stroke at age seventy-three. She lived to see the confirmation of her theory before she died. Looking back at her career, Margulis summed up her approach to controversy with a simple phrase that served as her mantra in decades of academic battles: "I don't consider my ideas controversial, I consider them right."

Creativity, a forceful personality, and technology changed how we view the history of life. Great revolutions happened when individuals combined to make ever more complex organisms, when formerly free-living creatures became parts of ever greater wholes. Every plant and animal on Earth today is an individual that contains a complex hierarchy of parts from organs to cells, organelles, and genes. The way this organization came about is a story spanning billions of years that begins near the origin of the planet itself.

Some Assembly Required

The deeper we venture into the past, the blurrier the picture of life becomes. Perhaps nobody knows this better than J. William Schopf, whose life's work has been to find evidence of the earliest living things on the planet. His hunt has taken him to the arid hillsides of Western Australia. It's a special place because the rocks are over three billion years old—among the oldest in the world. Accordingly, scientists have converged here to understand the workings of the early Earth. Such rocks have generally seen it all—they have been heated, squeezed, and heaved about in the eons since they were first deposited. Whatever originally lay inside, including fossils, is usually baked or crushed away.

Exploring a rock formation known as the Apex Chert in the early 1980s, Schopf noticed some rocks that seemed relatively undeformed for their age. Rocks that have been heated to high temperatures or submitted to high pressures contain characteristic minerals inside that formed as a consequence of this deformation. The Apex Chert had relatively few of those minerals. Knowing they were likely a rarity, Schopf brought the rocks to the lab to probe what was inside. Chert, a rock formed from the ooze on the seafloor, often contains the remains of creatures that settled to the bottom of the ocean after they died.

Working with cherts can be exacting. Each rock is sliced with a diamond saw, and the slivers are placed on a slide under a microscope for analysis. Schopf put two graduate students on the project, but, after dedicating a couple of years of long hours under the microscope, they found nothing. Picking up on their work, a third student looked for a few months and found some microscopic filaments inside the rocks. Thinking they were unremarkable, he put them in a specimen cabinet for later analysis. The student ended up taking a job in industry, and the specimens sat in the cabinet for two more years.

One day, not knowing what he had, Schopf pulled the cherts out of the cabinet for study. Some of the microscopic filaments looked like little slivers, strips, and ribbons. Most were set like a string of pearls, small circular structures attached to one another. Schopf had seen these patterns before, in living blue-green algae that form small colonies. But these cell-like structures came from rocks that were almost three and a half billion years old. Schopf made the bold announcement that he had found the earliest fossils on Earth, coming from rocks that had formed one billion years after the origin of the planet and the solar system.

Not everybody was convinced; along with the fanfare came vocal detractors. One critique was that structures like Schopf's filaments could be a natural outcome of the way the rock had formed over billions of years. The detractors claimed that the fragments were not fossils but a type of graphite produced by rocks crushed under high pressures. Journals were filled with papers arguing the pros and cons of Schopf's claim. Schopf even had a highly public debate with a prominent opponent. The subject, microscopic filaments inside rock, may seem painfully esoteric, but the issue at stake, understanding the earliest living things, was most definitely not.

Schopf tried another tack. Instead of comparing the shapes of the filaments and blue-green algae, he sought another clue about early life. A few decades after his original discovery, new technologies allowed scientists to look at the chemistry of the grains inside the rock and the putative fossils. The element carbon exists in several forms on the planet, and some kinds of carbon atoms are heavier than others. Living things metabolize carbon and preferentially use one type of it. Given this chemical specificity, life leaves a fingerprint in rocks based on the ratios of the different carbons inside.

Using a mass spectrometer, a machine about the size of a household dishwasher, Schopf and his colleagues probed the carbon content of the grains in the rock and those in the filaments. The filaments had the carbon signature of life. What's more, they represented at least five different kinds of living things. Some had the carbon fingerprint of creatures that had a primitive form of photosynthesis. Others looked like microbes known to metabolize methane as fuel. If the Apex Chert was a tiny window into the ancient Earth, it was showing that by three and a half billion years ago, life on the planet was already diverse.

We know rocks can be probed for chemical evidence of life. Even if the fossils are long gone, the chemical signature of life could remain. If creatures were metabolizing carbon, then the altered carbon content should lie like a residue in the rock. Probing the rocks of East Greenland for their carbon, a team from Yale found evidence of life in rocks even older than the Apex Chert. They were 4 billion years old, dating to 500 million years after the formation of the planet and the solar system.

What these inquiries show is that from these early beginnings until two billion years ago, the Earth was populated solely by single-celled creatures living alone or in colonies. The genes of each individual microbe gave rise to successive generations—one individual split into daughters, the daughters split, and the generations grew over time. Invention was mostly about developing new kinds of metabolism, chemical adaptations to more efficiently process energy, fuel, and wastes. Some species derived energy from sulfur or nitrogen, others from light and carbon dioxide. Still others utilized oxygen in processing energy. These single-celled creatures set the stage for revolutions to come.

Microbial metabolisms changed the world. For almost two billion years, blue-green algae were the most abundant living things on the planet. With photosynthesis, they used the light of the sun and carbon dioxide to make usable energy. Their waste product was oxygen. Blue-green algae exist as colonies, either in strips such as those Schopf found or in toadstool-shaped communities that could get as big as a microwave oven. Starting three and a half billion years ago, these colonies were abundant around the globe. By pumping oxygen into the air for billions of years, they fundamentally changed the atmosphere. Starting from an atmosphere with very little oxygen four billion years ago, oxygen levels increased to be able to support diverse kinds of life.

The rise in oxygen was a mixed blessing for microbes. For some, oxygen was a poison, whereas for others it opened up new possibilities. One type of microbe started to flourish—not surprisingly, one that could derive its energy from oxygen.

For billions of years, single-celled creatures were like a body without organs; they had no organelles with specialized functions inside them. Signs of change were first seen in fossils recovered from an iron mine in Ishpeming, Michigan, in 1992. These fossils look like coiled strips of cells and are about three and a half inches long. Coming from rocks almost two billion years old, they have the classic structure of a complex cell with organelles. As first glance they did not look the part, but these coiled strips heralded a revolution.

When a bacterium that metabolized oxygen teamed up with another microbe, a new kind of individual emerged on the planet. As Margulis showed, the merger was not one plus one equals two; it was more like one plus one equals four hundred. The host for this merger was a cell that had a nucleus and the machinery to generate different kinds of proteins. By incorporating an oxygen-consuming bacterium and converting it to be its own powerhouse, the new combined cell had the resources to make ever more complex proteins and behave in new ways.

No longer was the single-celled bacterium free to live on its own; it was part of a greater whole, a new, more complex individual with different parts. The formerly free-living bacterium could no longer reproduce by itself when needed; its functions were at the service of the host cell. And the new combined cell, now with the energy to live a more active existence and the machinery to make ever new kinds of proteins, became the harbinger for yet another significant change in the history of life.

The new cells, the supercharged protein factories, set the world up for the rise of yet another new kind of individual.

Coming Together Again

Every animal and plant on Earth has a body composed of many cells: recall that the worm *C. elegans* has about a thousand cells, while humans have four trillion. Despite large differences in the number of cells, bodies share very deep and ancient similarities.

The earliest bodies in the fossil record do not look like much. Found in rocks over 600 million years old from Australia, Namibia, and Greenland, they are mere impressions. Whatever was inside the rock has long eroded away. Ranging in size from a nickel to a dinner plate, they look like ribbons, fronds, or disks. While the shapes are not inspiring, how they arose is another matter. These are the earliest fossils of multicellular life, creatures with bodies. And bodies were themselves an entirely new kind of individual on Planet Earth.

Philosophers have various definitions of what an individual is, but in the most basic sense, individuals have a beginning and an end, a birth and death, and can reproduce; importantly, the different parts inside them work together to make a functioning whole. Each of us is an individual because our body, like the bodies of other plants and animals, has all of these properties. Moreover, our bodies remain healthy only because their constituent parts work together to make larger entities. For example, trillions of nerve cells make brains, but a list of them would never tell how thoughts, feelings, and memories form. Brains can produce thoughts, while individual neurons cannot—thinking is a higher-order property that comes from the organization of billions of nerve cells.

The diverse cells within bodies are individuals, too, but in a different way. Each cell has a birth and a death. Each cell reproduces. And each cell has parts inside that interact. But consider: a human body contains nearly four trillion cells. Those cells form organs, each having its own size, shape, and position in the body. Cells need to reproduce and die in a regular way in order to make the heart, liver, and intestines the correct size in the proper place in the body. The coordination of cells is what makes a body possible. Cells do not behave individually; their growth, death, and life are regulated to make a working body. By limiting their reproduction and dying at the right time, cells inside bodies sacrifice themselves for a higher good, the functioning of the body as a whole.

A special molecular machinery gives cells the ability to work together and make bodies. Different cells have to be able to stick together. It would be challenging to have a solid body in which the cells did not adhere to one another in very precise ways. Skin cells, for example, have special mechanical properties that allow them to attach to one another to make sheets of tissue. They make the collagens, keratins, and other proteins that give the tissue its characteristic feel. Finally, cells in bodies need ways to communicate with one another, to coordinate their reproduction, death, and gene activity. And again, proteins are the way this happens: different proteins convey messages to cells that tell them where and when to divide, die, or secrete more proteins.

The genetic machinery that makes this possible is the gene families we discussed in Chapter 5. Each gene in the family makes a protein that is subtly different from its cousins. For example, one class of protein, cadherins, resides in one hundred different kinds of cells, each specific to a different kind of tissue—skin, nerve, bone, and so forth. These proteins both hold cells together, as in the skin, and serve as a way for cells to communicate chemically, telling one another when to divide, die, or make other proteins.

Here's the important part: these proteins are expensive for a cell to manufacture, because synthesizing and assembling them requires a significant amount of metabolic energy. That is the reason bodies could not have originated without Margulis's new kind of cell. The merger she envisioned brought together a powerhouse and a protein maker. This chimera of a cell now had the energy and DNA to make the diversity of proteins that enabled the evolution of bodies. It could attach to other cells, communicate with them, and behave in novel ways.

Over the course of billions of years, we have witnessed the succession of ever more complex individuals: the origin of one new kind of individual, a cell with organelles, enabled the origin of the next, a body with many cells.

This sequence raises the question, How did bodies arise?

My colleague Nicole King at Berkeley has spent her career studying one special kind of single-celled creature. Microscopic and shaped like a jelly bean, it has one unusual feature: a circle of hairs projects straight out from one end like a frightened monk's tonsure. Choanoflagellates, or choanos, as King affectionally calls them, have special features. Their genome was sequenced a decade ago and compared to both animals and other single-celled creatures. The result was the realization that choanos are the closest relative of multicellular animals. This relationship means that they might provide clues to the mechanisms behind the origin of bodies.

Moreover, choanos perform an important trick. For most of their lives they swim freely, feathering their hairs to move about. Then, at special times, a trigger goes off, and they combine to form clumps. Known as rosettes for their flower-like shape, these clumps can have ten or more formerly separate choanos attached to one another. The transition from single-celled creature to a clump of many cells, something that took billions of years in evolution, happens in an instant in choanos.

King may have trained as a molecular biologist, but she thinks like a paleontologist. Just as fossil hunters look at living crea-

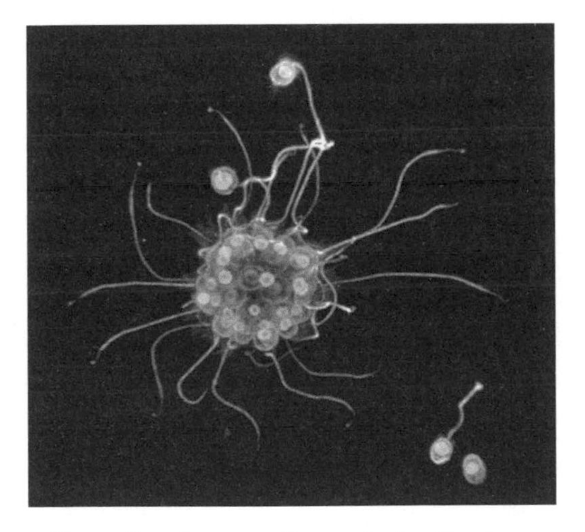

Choanoflagellates can form colonies, like the one depicted here.

tures and ask what their antecedents might have been, King does the same with the processes that form bodies, asking, What molecular mechanisms are necessary to build bodies, and where did they come from?

If, as we saw, cells manufacture special proteins to make bodies, then clues to the origin of bodies would come from exploring how those molecules originate. Genomes now hold the answers, with sequences of choanos, bacteria, and diverse microbes ripe for exploration. Using computer databases, scientists can look at a genome of a creature and know precisely what proteins it can make.

When the genome of choanos was sequenced, it revealed one incredible fact. Many of the proteins that build bodies are already present in this single-celled creature. They use the proteins to form rosettes or find and consume prey. This observation set King and others on an even broader hunt, to look at the genomes of diverse microbes. The result is a pattern of evolution we have seen before.

King and her colleagues discovered that versions of the proteins that animals use to build bodies, such as collagens, cadherins, and many others, are present in a menagerie of single-celled creatures, from bacteria to more complex ones with organelles. What do they do with these proteins if they are not making bodies? They use them to attach to prey or to parts of their environment. They use them to avoid predators. Single-celled creatures also can communicate with one another using chemical cues. Microbes adapting to their world developed the chemical precursors that animals later used to make bodies. Multicellular life is possible only because new combinations of molecules were repurposed from their original function in single-celled life. The great inventions that made bodies possible predate the origin of bodies themselves.

King recently discovered the trigger for the formation of a choanoflagellate rosette. When choanos find themselves in the presence of a particular bacterial species, they start to make proteins that cause them to clump. We don't know exactly why the bacterium does the triggering. It may well be that it has a chemical signal that stimulates clumping behavior. But the observation is intriguing: not only did single-celled creatures provide the raw material for bodies, they may have induced them as well.

It took both potential and opportunity for bodies to emerge. The machinery needed to make bodies was around for eons before bodies first appear in the fossil record. By one billion years ago, oxygen had made a new world for creatures that were prepared to thrive in it. With the rise in the levels of oxygen in the atmosphere, creatures that metabolized oxygen could live a higher-energy lifestyle. That energy was put to use with Margulis's new kind of cell. The ability to manufacture proteins on the industrial scale needed to make bodies is possible only because the cells had a powerhouse that was fueled by oxygen. And there was fuel aplenty by one billion years ago.

Sum of the Parts

The organization of bodies is much like Russian dolls: bodies contain organs that are composed of tissues that are made of cells that have organelles, all of which have genes inside. Over billions of years of evolutionary time, different parts essentially relinquished their individuality to become parts of greater wholes. Free-living microbes combined to make a new kind of cell. That new cell had special properties that allowed for yet another new combination, multicellular bodies. Successively more complex kinds of individuals have emerged with ever more intricate parts.

Bodies and cells rely on highly controlled behaviors of their constituent parts. But beneath that order lies cacophony. Coordinating parts in a body means corralling the competing interests that lie in the different cells and parts of the genome. Different genes, organelles, and cells inside bodies continually reproduce. Left unchecked, one part can take over. The conflict between parts that behave selfishly and try to reproduce unchecked, and the needs of the body is a story of health, disease, and evolution. The outcome can be a mother of invention or a path to catastrophe.

Imagine a cell that behaves on its own and simply divides and reproduces with wild abandon, or, conversely, does not die at the right time or place. Cells like this can take over the body and break it down. In fact, this is precisely what cancer does: cancerous cells break the rules and function selfishly, coordinating neither their reproduction nor their death with the needs of the individual in which they reside.

Cancer reveals an essential tension between parts and wholes—in our case, between the components that make bodies and the bodies themselves. If parts behave in their own short-term self-interest and divide unchecked, they can lead the body to break down. Cancer is a disease of genetic mutations that accumulate and cause cells to proliferate too rapidly or to not die properly. In response, bodies have developed defenses, such as immune responses, that pick off misbehaving cells. When these checkpoints and defenses ultimately break down and the cells' behavior becomes uncontrollable, cancer turns deadly.

A similar conflict takes place inside the genome. Barbara McClintock's jumping genes exist to make copies of themselves, much like a cancer cell does. The war inside is between rogue selfish elements that want to proliferate wildly and the individual organism. With genes struggling to contain selfish elements, viruses continually invading, and trillions of cells working together to keep bodies functioning, multicellular bodies are a confederation of parts that arose at different times, sometimes in different places. These parts, some in conflict, some cooperating, all changing over time, fuel the fire of evolution. Bodies can evolve and vary in new ways because of the diversity of parts and the ways they interact.

Mixology

Wheels have existed on Planet Earth for about six thousand years. Suitcases have been around for centuries. Suitcases with wheels were invented a few decades ago and changed life for many who travel. Every time I am in an airport, I celebrate how a revolutionary invention can come from finding a new combination.

Margulis's organelles revealed the power of combination as a source of invention in the natural world. What if a lineage doesn't invent something itself but instead acquires a feature that arose in another species? The mitochondria that power our cells were not invented by changes to our own genome, when our ancestor was a single-celled creature. They were invented elsewhere, then taken in and reused as those ancient bacteria merged with our lineage. Similarly, viruses, through millions of years of infecting genomes, brought them the capability to make new proteins. When those viruses were repurposed, new molecules to aid in pregnancy and memories came about.

Traits can appear in one species only to be borrowed, stolen, and modified for new uses by another. Hosts can inherit a ready-made invention rather than having to build it themselves. Combinations of parts, and the new kinds of individuals that can emerge from them, can open up evolutionary opportunities.

For billions of years, life existed as single cells, and the inventions occurred in the ways creatures metabolized the energy and chemicals around them. Life was small. The emergence of ever more complex individuals brought new ways of making proteins, moving about, and feeding. Creatures with bodies—animals, plants, and fungi—are relative newcomers to the planet, and they are all composed of cells derived from the merger of different individuals. The advent of bodies opened up a new way of evolving. Creatures made of many cells, each powered by organelles, could get big and develop new tissues and organs. The results are the diversity of tissues and organs that help animals fly to the highest altitudes, swim at the bottom of the ocean, and devise satellites to probe the far reaches of the solar system.

Appropriating the Future

Combining, borrowing, and repurposing technologies and inventions from other species has been our multibillion-year past. It is also part of our future.

In 1993, the Spanish microbiologist Francisco Mojica was studying salt marshes in Costa Blanca, in southern Spain. His goal was to understand how bacteria evolved to thrive in an extremely salty habitat. Something in their genome was giving them resistance to an environment that is deadly for most species. In almost a decade of following a trail of discovery, he sequenced their genome and uncovered a puzzling feature. Most of their DNA had a standard bacterial sequence of different letters. But a small number of places had a short stretch that formed a palindrome, reading the same way backward and forward, like the name Hannah, only in this case with the letters A, T, G, and C. Moreover, one short block of palindromes would be spaced evenly from another one, forming a repeating pattern: palindrome, space of other sequences, palindrome, and another space of sequences. In fact, in an example of a multiple in science, a Japanese laboratory had identified these palindromic sequences about a decade before.

Thinking this no random occurrence, Mojica searched other bacteria for this strange pattern. Lo and behold, he found it to be exceedingly common, occurring in more than twenty species. Such a well-defined and widespread genomic pattern must have a function, but what could it be?

By this point Mojica had started his own lab in Spain, but he lacked enough money to do sequencing or any high-tech lab work. Undeterred, he used his desktop PC, some word-processing software, and an Internet connection to a gene database. He input the sequence of palindromes and the spaces of sequence that separate them to see where else they might reside. He found hits, but they weren't in other bacteria. The absolute best match for them was in a virus. Moreover, the virus was one to which this species of bacterium had developed resistance. He plodded on, looking at eighty-eight spacer regions that separate the palindromes. More than two-thirds of them corresponded to viruses to which the bacterium was resistant. It was almost as if these regions were protecting the bacterium from viral invasion.

Mojica made a bold and untested hypothesis—that this palindrome-space system is a bacterial weapon against viruses. He wrote up his idea and submitted it to some leading journals. One rejected it without even sending it out for peer review. Another sent it back for lacking "novelty or importance." This process was repeated five times until the work ended up in a molecular evolution journal. In the same year a laboratory in France, using slightly different methods, published the same idea independently.

Then a network of other laboratories got into the hunt. A bacterial defense would be a boon to the yogurt industry, whose cultures suffer at the hands of viral invaders. With this incentive, it soon was convincingly demonstrated that this system evolved in an arms race with viruses. Viruses attack bacteria as well as humans. We fend most of them off with our immune system. This bacterial mechanism confers on bacteria a kind of immunity. It uses a molecular guide and scalpel: the palindromes help form the guides that bring a molecular scalpel to cut the viral DNA to render it harmless. It is a defense against the viruses' selfish nature to infect, divide, and take over other genomes.

Following these discoveries, a number of laboratories around the world did creative and groundbreaking research on the molecular scalpel (known as *Cas9*) to show how it is possible to repurpose this system to edit not just viral DNA but the DNA in any creature. Papers were submitted to scientific journals within months of one another describing ways to modify the bacterial system for use in other species. The technique, known as CRISPR-Cas (which we saw Nipam Patel use to move appendages about in *Parhyale*), is the basis for genome editing, a now-familiar mechanism that can edit the genomes of plants, animals, and people for benefits in everything from agriculture to health. And it is only the beginning: refined techniques that are more precise, rapid, and efficient are being developed almost monthly.

This technique can rewrite parts of the genome practically overnight. In evolutionary history, these kinds of changes have taken millions of years to happen. While it is still the early days for the technology and news is often hyped, it is clear that we can rewrite parts of the genome of plants and animals quickly and cheaply. My lab has applied this technique to fish, using the crudest application: deleting genes. Other labs are able to cut and paste entire sections of the genome, moving genes and their switches from one species to another or from one individual to the next.

The discovery of CRISPR-Cas genome editing follows a well-worn path of four billion years of evolutionary invention. The breakthrough that led to the technological revolution happened not in the place we associate it with, genome editing in animals and plants, but in a different place—understanding saltwater ecosystems. What followed was a tangled path of discovery, with multiple inventors developing similar ideas at the same time, combining technologies, and breathing the same air of discovery. And just as in biological inventions, a key moment came from repurposing an invention in one species, bacteria, for use by another, ourselves. The development of CRISPR-Cas involved hundreds of senior and junior scientists working in parallel. The quirks of history, multiples, and numerous unexpected antecedents make this story perfect for one species—lawyers. Patent battles are at the heart of deciphering the history of CRISPR-Cas.

There is something sublime to the notion that our conscious brain has achieved what cells and genomes have been doing on their own for billions of years. A technology invented in one creature, bacteria, has been taken, modified, and co-opted to change others. The brain that appropriated and modified these biological inventions is itself composed, in part, of repurposed viral proteins and is powered by formerly free-living bacteria. New combinations can change the world.

EPILOGUE

On Christmas Day 2018, I had been holed up in my tent most of the morning because of a summer blizzard. As the weather cleared, I climbed a ridge above camp to stretch my legs. Feeling increasingly liberated with each passing step, I eventually found myself at the summit of Mount Ritchie, one of the ridges in the Transantarctic Range of Antarctica. I was surrounded by a plateau of ice larger than the continental United States. Our team had moved our fossil hunt to rocks older than those that held the fishapod *Tiktaalik roseae* near the North Pole. Here at the opposite end of the planet, our search was for some of the earliest fish with bony skeletons. Rocks of the right type and age in which such fossils might be found brought us to the mountains in this part of Antarctica.

Here, the mountaintops poke through the glaciers, exposing a layer cake of colors that form a vibrant contrast to the sea of whiteness around them. Layer after layer of reds, browns, and greens hold 400 million years of the history of life and the planet. The structures inside the rocks show that this polar region was once a giant tropical delta the size of the Amazon and, later, a place of intense volcanic activity. Life has changed here too. The rocks at the bottom are almost 400 million years old and contain mostly fish, while those at the top are 200 million years old and hold ecosystems with a diversity of reptiles.

Seen from this distance, it is tempting to look at these layers and envision an orderly progression of evolutionary change. At this scale more globally, layers with the first microbes lie beneath those with the earliest animals, those with earliest fish lie beneath those with amphibians, those with the earliest amphibians lie beneath those with reptiles, and so on.

We tend to fill the gaps in our knowledge with our own biases, usually some combination of hope, expectation, or fear. Our minds have a tendency to connect the dots of past events to construct a narrative in which one change leads to the next in a linear sequence. We've all seen cartoons of human evolution that show a parade that extends from monkeys to apes to humans going from hunched creatures on four legs to those walking on two. Often this depiction is satirical, with the end of evolution being a human on the couch watching *The Simpsons* or glued to their phone. That view of history is deeply entrenched. How many times have you heard the term *missing link*, as if there were a great chain of evolution in which one link leads inexorably to the next? Or that missing links should look like an exact blend of the traits of ancestors and their descendants?

True, the first fish appear before the first land-living creatures in the fossil record. But as we have seen, the more we look at fossils, embryos, and the DNA of diverse species, the more we find that many of the changes that allowed animals to live on land arose earlier, while fish were living in water. Every major revolution in the history of life followed the same path. Nothing ever begins when we think it does: antecedents appear earlier and in different places than we imagine. And as Darwin knew when he responded to St. George Jackson Mivart more than 150 years ago, the history of life couldn't have happened any other way.

Darwin didn't know about DNA, or the workings of the cell,

or the ways genetic recipes build bodies during embryological development. Ever twisting, turning, and at war with itself and external invaders, DNA provides the fuel for evolution's changes. Ten percent of our genome is made up of ancient viruses, and at least another 60 percent consists of repeated elements made by jumping genes gone wild. Only 2 percent is made up of our own genes. With cells and genetic material of different species merging and genes continually duplicating and repurposing, life's history flows more like a braided and meandering river than a straight channel. Mother Nature is like a lazy baker who crafts a bewildering variety of concoctions by repurposing, copying, modifying, and redeploying ancient recipes and ingredients. In this way, through eons of jury-rigging, duplicating, and co-opting, single-celled microbes have evolved to the point where their descendants thrive in every habitat on the planet and have even walked on the moon.

Every now and then I return to the diagram that launched my career three decades ago: the image of a fish connected to an amphibian by an arrow. It now seems quaint, even naïve. The figure captured evolutionary biology at a time before we knew much about genomes, viral invaders, or the genes that build bodies. We didn't know about the limbed fish that my colleagues and I would discover in 2004, nor about any of the other recently uncovered fossils that tell us of other major events in life's history. Today we are doing science that we could not have dreamed of only a few decades ago. Like the history of life, scientific discovery is full of unexpected twists, turns, dead ends, and opportunities that change the way we see the world around us. The ideas we use to probe nature's diversity are themselves repurposed and modified from those our predecessors developed decades, if not centuries, ago.

The poet William Blake wrote of seeing "the universe in a grain of sand and heaven in a wildflower." When you know how to look, you can see billions of years inside the organs, cells, and DNA in all living things and relish our connections to the rest of life on the planet.

FURTHER READING AND NOTES

A number of excellent general introductions to the history of life and the planet are available. Richard Fortey, an accomplished paleontologist and a gifted writer, has produced two books with a broad sweep: *Life: A Natural History of the First Four Billion Years of Life on Earth* (New York: Vintage, 1999) and *Earth: An Intimate History* (New York: Vintage, 2005). Richard Dawkins worked through the tree of life in reverse order, then narrated how species have changed over time and described the tools we use to reconstruct that history in *The Ancestor's Tale: A Pilgrimage to the Dawn of Evolution* (New York: Mariner Books, 2016). Compelling and informative resources on life's earliest history include Andrew Knoll, *Life on a Young Planet: The First Three Billion Years of Evolution on Earth* (Princeton, NJ: Princeton University Press, 2004), Nick Lane, *The Vital Question: Energy, Evolution, and the Origins of Complex Life* (New York: Norton, 2015); and J. William Schopf, *Cradle of Life: The Discovery of Earth's Earliest Fossils* (Princeton, NJ: Princeton University Press, 1999). For a lively and comprehensive history of the fossil record, see Brian Switek, *Written in Stone: Evolution, the Fossil Record, and Our Place in Nature* (New York: Bellvue Literary Press, 2010).

In the past few years a number of excellent general books on genetics and heredity have appeared, almost like multiples in the evolutionary record: Siddhartha Mukherjee, *The Gene: An Intimate History* (New York: Scribner, 2017); Adam Rutherford, *A Brief History of Everyone Who Ever Lived: The Human Story Retold Through Our Genes* (New York: The Experiment, 2017); and Carl Zimmer, *She Has Her Mother's Laugh: The*

Powers, Perversions, and Potential of Heredity (New York: Dutton, 2018). For a gripping account of molecular evolution and many of the new ideas generated by it, see David Quammen, *The Tangled Tree: A Radical New History of Life* (New York: Simon and Schuster, 2018).

PROLOGUE

References for "fish with arms, snakes with legs, and apes that can walk on two legs" include N. Shubin et al., "The Pectoral Fin of *Tiktaalik roseae* and the Origin of the Tetrapod Limb," *Nature* 440 (2006): 764–71; D. Martill et al., "A Four-Legged Snake from the Early Cretaceous of Gondwana," *Science* 349 (2015): 416–19; and T. D. White et al., "Neither Chimpanzee nor Human, *Ardipithecus* Reveals the Surprising Ancestry of Both," *Proceedings of the National Academy of Sciences* 112 (2015): 4877–84.

1. FIVE WORDS

The seminar was taught by the late Farish A. Jenkins, Jr., who became a mentor of mine and a collaborator on the expeditions that led to the discovery of *Tiktaalik roseae*. The diagram that inspired me made its way into a fabulous little book on great transformations in vertebrate evolution: Leonard Radinsky, *The Evolution of Vertebrate Design* (Chicago: University of Chicago Press, 1987), figure 9.1, p. 78. Farish was close friends with Radinsky, who had shared drafts of the book's illustrations, done by Sharon Emerson, with him for the course. Coincidentally, Radinsky was my predecessor as chair of the anatomy department at the University of Chicago. Little could I have known in graduate school that his diagram would inspire me to follow in his footsteps decades later.

Lillian Hellman's quote appears in her autobiography, *An Unfinished Woman: A Memoir* (New York: Penguin, 1972). The biological translation for the concepts she expressed are *exaptation* and *preadaptation*. The subtle distinctions between them are discussed in Stephen J. Gould and Elisabeth Vrba, "Exaptation—A Missing Term in the Science of Form," *Paleobiology* 8 (1982): 4–15. See also W. J. Bock, "Preadaptation and Multiple Evolutionary Pathways," *Evolution* 13 (1959): 194–211. Both important papers contain numerous examples.

My history of St. George Jackson Mivart is taken from J. W. Gruber, *A Conscience in Conflict: The Life of St. George Jackson Mivart* (New York: Temple University Publications, Columbia University Press, 1960). Mivart's *On the Genesis of Species*, published in 1871, is now available online at https://archive.org/details/a593007300mivauoft.

The sixth edition of Darwin's *On the Origin of Species* is also available online, at https://www.gutenberg.org/files/2009/2009-h/2009-h.htm.

Gould's take on "the 2% of a wing problem" is in Stephen Jay Gould, "Not Necessarily a Wing," *Natural History* (October 1985).

My account of Saint-Hilaire's life and work is derived from H. Le Guyader, *Geoffroy Saint-Hilaire: A Visionary Naturalist* (Chicago: University of Chicago Press, 2004), and from P. Humphries, "Blind Ambition: Geoffroy St-Hilaire's Theory of Everything," *Endeavor* 31 (2007): 134–39.

The original description of the Australian lungfish is in A. Gunther, "Description of *Ceratodus*, a Genus of Ganoid Fishes, Recently Discovered in Rivers of Queensland, Australia," *Philosophical Transactions of the Royal Society of London* 161 (1870–71): 377–79. The history of the discovery is in A. Kemp, "The Biology of the Australian Lungfish, *Neoceratodus forsteri* (Krefft, 1870)," *Journal of Morphology Supplement* 1 (1986): 181–98.

On the developmental and evolutionary relationships between swim bladders and lungs, see Bashford Dean, *Fishes, Living and Fossil* (New York: Macmillan, 1895). His catalog of the armor collection at the Metropolitan Museum of Art is available digitally at http://libmma.contentdm.oclc.org/cdm/ref/collection/p15324coll10/id/17498. For a synopsis of his work and life, see https://hyperallergic.com/102513/the-eccentric-fish-enthusiast-who-brought-armor-to-the-met/.

Analyses of air breathing include K. F. Liem, "Form and Function of Lungs: The Evolution of Air Breathing Mechanisms," *American Zoologist* 28 (1988): 739–59; and Jeffrey B. Graham, *Air-Breathing Fishes* (San Diego: Academic Press, 1997). Both show how lungs are the primitive condition for bony fish and corroborate the comparison between swim bladders and lungs.

Recent genetic comparisons between lungs and swim bladders have found deep similarities. See A. N. Cass et al., "Expression of a Lung Developmental Cassette in the Adult and Developing Zebrafish Swimbladder," *Evolution and Development* 15 (2013): 119–32. Dean and his contemporaries would be proud.

The story of lungs is only one exemplar of the importance of a change in function at the origin of land-living fish.

Gunnar Säve-Söderbergh, at the age of twenty-two, was in charge of a small team of geologists exploring the rocks in the region for fossils. The hunt was a relatively simple and low-tech affair. Each day the team would disperse across the rocks and look for bones weathering out on the surface. When they found some, they would trace the fragments to attempt to identify the rock layer they came from. These were precisely the techniques that my team would use almost eighty years later in the Canadian Arctic to find the fishapod *Tiktaalik roseae*.

Säve-Söderbergh's hunt was for the earliest creatures to walk on land. At the time, nobody had ever found a hint of limbed animals in Devonian-age rocks, which are about 365 million years old. His goal was to go to more ancient rocks to find a fishlike amphibian, a species that blurred the distinction between fish and amphibian.

Säve-Söderbergh was legendary for his energy; he'd work late nights and hike enormous distances to find fossils. He was also supremely confident. Pessimists don't find fossils; you have to believe that there are fossils in the rocks to devote the long hours and many failed efforts required to find them. Each day his team were to place their findings in one of two boxes: P for fish (Pisces) and A for amphibians. It was a bold move. Nobody had ever found an amphibian in rocks of this age. As you can imagine, over the course of the 1929 field season, the fish box burgeoned with fossils and the amphibian box remained empty.

Near the end of the season, Säve-Söderbergh found a number of odd-looking fragments of bone in the rubble of Celsius Berg, a deep red butte adjacent to the ice of the East Greenland Sea. He collected nearly a dozen plates of bone, each of which was embedded in rock obscuring most of its structure. With their bumps and ridges, these plates looked like some of the fossil fish known at the time. Judging from what was preserved, they belonged in the fish box. They were clearly from a skull but were too flat to be associated with any fish known at the time. Säve-Söderbergh thought they might be amphibian. Ever the optimist, he threw them into the A box.

Returning to Sweden, Säve-Söderbergh began the laborious process of removing grains from the rock that surrounded each bone. Removing the layers revealed a true marvel. He had found what looked like a fish in body shape, but its head had the long snout and flat shape of an amphibian. Säve-Söderbergh had found his early amphibian.

The fossil became a celebrity. Säve-Söderbergh would have become one, too, but he died tragically from tuberculosis before his thirtieth birthday.

The story of Säve-Söderbergh's work was told by a colleague and friend of his, Erik Jarvik. Jarvik, a member of the early expeditions, included a brief history of the Greenlandic expeditions in his hefty monograph on *Ichthyostega*, one of the first discovered Devonian tetrapods: E. Jarvik, "The Devonian Tetrapod *Ichthyostega*," *Fossils and Strata* 40 (1996): 1–212. Carl Zimmer, *At the Water's Edge: Fish with Fingers,*

Whales with Legs (New York: Atria, 1999), discusses Säve-Söderbergh, Jarvik, and the larger history of the field in a highly readable account.

Five decades after Säve-Söderbergh, my colleague Jenny Clack, from Cambridge University, returned to Celsius Berg and his other sites to look with new eyes. Her team of paleontologists were well-versed in Säve-Söderbergh's discoveries and notes. Their goal was to find missing parts of the skeleton, the ones that he did not collect. Lost in all the hoopla around the fossils was the fact that their limbs were poorly known. Hitting the rocks, Clack set out to correct that. With the right team, good weather, and the knowledge that the rocks held promise, she came back with a trove of fossils. And these fossils had well-preserved limb skeletons connected to them.

The limbs had the classic one bone–two bones–little bones–digits pattern seen in everything with limbs, whether a mammal, bird, amphibian, or reptile (see pages 119–20). The surprise lay in the hand and foot. These animals had more than five fingers and toes; they had as many as eight. The extra digits made the limbs broad and flat. Everything about them, from their proportions to the muscle scars on individual bones, implies that they were used as paddles or oars in water. The entire limb was more like a flipper than a hand.

What does this have to do with Darwin's five words? The earliest animals possessing limbs with fingers and toes used them not to walk on land but to paddle about in water or maneuver through the shallows of swamps and streams. As with lungs, the earliest uses of these great inventions of land-living creatures was not to live on land but to make use of an aquatic environment in new ways. The organ arose early in one setting, with the big revolution—the shift to a new environment—coming about from repurposing it for a new function.

Clack's magisterial *Gaining Ground: The Origin and Evolution of Tetrapods* (Bloomington: Indiana University Press, 2012) is the result of a lifetime of work on the origin of tetrapods by a person who brought that field into the modern age. Her book includes both the science and the history of the field along with an important personal account of her work in the Devonian sites in Greenland.

In animals both living and long extinct, lungs, arms, elbows, and wrists all first appear in aquatic animals. The major revolution from life in water to life on land didn't involve new inventions. It involved changes in inventions that had come about millions of years before.

If history were a single path of change, where one step led inexorably to the next, each with a gradual improvement for a single function, major changes would be impossible. Every major transition would require waiting for not just one invention but a whole patent agency full of them to arise in concert. If, on the other hand, the inventions are already there, doing something else, a simple repurposing can open up new pathways of change. This capability for change is the power of Darwin's five words.

Knowing that ancient creatures lived in water with lungs, arm bones, wrists, and even digits, our question about the invasion of land by fish changes. Instead of "How could creatures ever evolve to walk on land?," the question becomes, "Why didn't the transition happen sooner in the history of the planet?"

Rocks again hold clues. For billions of years, all of the rocks on Planet Earth lacked one thing. Rocks from 4 billion to about 400 million years ago hold evidence of vast oceans and smaller seaways, and on land, fast rivers capable of moving boulders and rocks. But, importantly, there was no evidence for plants on land.

Imagine a world without plants on land. Plants decay when they die and create soil. Plants have roots that hold soil together. This was a barren, rocky world lacking soil. It also lacked any food that animals could eat.

Land plants first appear in the fossil record about 400 million years ago, and insect-like creatures soon thereafter. The invasion of land by plants created a whole new world, one where bugs and insects could thrive. Some of the fossil plant leaves show damage, implying that they were eaten by these early bugs. With plants on land came detritus as they died and rotted. The resulting soils made possible shallow streams and ponds to serve as habitats for fish and amphibians.

The reason fish with lungs didn't move to land earlier than 375 million years ago was that it was inhospitable until then. Plants, and the insects that followed them, changed everything; ecosystems now were habitable for any fish with the ability to spend short periods on land. Only when new environments appeared could our distant fish ancestors take those first steps, using organs that had already appeared while they were in water. Timing is everything.

Recent geological studies have shown how plants have changed the world, most notably how the invasion of land by plants changed the nature of the streams that existed in the Devonian. Plants with roots allow the formation of soils to form stable banks for shallow streams. For further discussion and analysis, see M. R. Gibling and N. S. Davies, "Palaeozoic Landscapes Shaped by Plant Evolution," *Nature Geoscience* 5 (2012): 99–105.

For general reviews of dinosaur evolution and bird relationships, and popular accounts by dinosaur scientists, see Lowell Dingus and Timothy Rowe, *The Mistaken Extinction* (New York: W. H. Freeman, 1998); Steve Brusatte, *The Rise and Fall of the Dinosaurs: A New History of a Lost World* (New York: HarperCollins, 2018); and Mark Norell and Mick Ellison, *Unearthing the Dragon* (New York: Pi Press, 2005).

For a lovely popular account of Huxley's work on *Archaeopteryx* and the origin of birds, see Riley Black, "Thomas Henry Huxley and the Dinobirds," *Smithsonian* (December 2010).

On Baron Nopcsa, his colorful life, and his pathbreaking science, see E. H. Colbert, *The Great Dinosaur Hunters and Their Discoveries* (New York: Dover, 1984); Vanessa Veselka, "History Forgot This Rogue Aristocrat Who Discovered Dinosaurs and Died Penniless," *Smithsonian* (July 2016); and David Weishampel and Wolf-Ernst Reif, "The Work of Franz Baron Nopcsa (1877–1933): Dinosaurs, Evolution, and Theoretical Tectonics," *Jahrbuch der Geologischen Anstalt* 127 (1984): 187–203.

John Ostrom's work was published in a number of papers in the 1960s and '70s, including his formal description of *Deinonychus:* J. Ostrom, "Osteology of *Deinonychus antirrhopus*, an Unusual Theropod from the Lower Cretaceous of Montana," *Bulletin of the Peabody Museum of Natural History* 30 (1969): 1–165. Papers that followed included J. Ostrom, "*Archaeopteryx* and the Origin of Birds," *Biological Journal of the Linnaean Society* 8 (1976): 91–182; and J. Ostrom, "The Ancestry of Birds," *Nature* 242 (1973): 136–39. For an account of Ostrom's contributions, see Richard Conniff, "The Man Who Saved the Dinosaurs," *Yale Alumni Magazine* (July 2014).

Recent surveys of the origin of features have spanned the fields of paleontology and developmental biology. See R. Prum and A. Brush, "Which Came First, the Feather or the Bird?," *Scientific American* 288 (2014): 84–93; and R. O. Prum, "Evolution of the Morphological Innovations of Feathers," *Journal of Experimental Zoology* 304B (2005): 570–79.

2. EMBRYONIC IDEAS

Duméril's story is best related by his initial surprise, then his ultimate solving of the puzzle. After doing so, he set up a breeding colony of axolotls and generously gave them away to any researcher who wanted them. Descendants of that population are likely in labs today. You would not know it from the title, but a good recent account of Duméril is G. Malacinski, "The Mexican Axolotl, *Ambystoma mexicanum:* Its Biology and Developmental Genetics, and Its Autonomous Cell-Lethal Genes," *American Zoologist* 18 (1978): 195–206. Some of Duméril's early work appeared in M. Auguste Duméril, "On the Development of the

Axolotl," *Annals and Magazine of Natural History* 17 (1866): 156–57; and "Experiments on the Axolotl," *Annals and Magazine of Natural History* 20 (1867): 446–49.

The field of embryology is blessed with some textbooks that are so good they have driven research in the field. These include Michael Barresi and Scott Gilbert, *Developmental Biology* (New York: Sinauer Associates, 2016); and Lewis Wolpert and Cheryll Tickle, *Principles of Development* (New York: Oxford University Press, 2010).

My treatment of von Baer (including his quote on misidentifying embryos in vials) and Pander is based in part on historical work by Robert Richards, available online at home.uchicago.edu/~rjr6/articles/von%20Baer.doc.

Stephen Jay Gould's *Ontogeny and Phylogeny* (Cambridge, MA: Belknap Press, 1985) has a wonderful history of embryology in the first half, where he covers the work of von Baer, Haeckel, and Duméril. A short review paper is a superb follow-up: B. K. Hall, "Balfour, Garstang and deBeer: The First Century of Evolutionary Embryology," *American Zoologist* 40 (2000): 718–28.

Over the years, while many learned Haeckel's ideas in school, scientists in the field had a love/hate reaction to him: some were acolytes of his work, while others, like Garstang, thought him a fraud. Recent histories have held a variety of views, as seen in Robert Richards, *The Tragic Sense of Life: Ernst Haeckel and the Struggle over Evolutionary Thought* (Chicago: University of Chicago Press, 2008). Some recent embryologists believe that some of Haeckel's original diagrams were, to put it charitably, drawn to emphasize his main points: M. K. Richardson et al., "Haeckel, Embryos and Evolution," *Science* 280 (1998): 983–85.

Apsley Cherry-Garrard, *The Worst Journey in the World* (London: Penguin Classics, 2006), is a classic of expedition literature. I read it before my first Antarctic expedition. It made McMurdo Sound, Hut

Point, and Mount Erebus feel like familiar landscapes when I saw them for the first time.

Walter Garstang, *Larval Forms and Other Zoological Verses* (Oxford: Blackwell, 1951), was republished by the University of Chicago Press in 1985.

Heterochrony has a vast literature stemming from the days of Garstang, if not before. Whole taxonomies of the rates and timing of development have been proposed. For a snapshot of some of the major approaches (with good references), see P. Alberch et al., "Size and Shape in Ontogeny and Phylogeny," *Paleobiology* 5 (1979): 296–317; Gavin DeBeer, *Embryos and Ancestors* (London: Clarendon Press, 1962); and Stephen Jay Gould, *Ontogeny and Phylogeny* (Cambridge, MA: Belknap Press, 1985). Gould's book had a large impact in the 1980s, leading to renewed interest in the approach.

Amphibian biology and metamorphosis are discussed in W. Duellman and L. Trueb, *Biology of Amphibians* (New York: McGraw-Hill, 1986); and D. Brown and L. Cai, "Amphibian Metamorphosis," *Developmental Biology* 306 (2007): 20–33. Duellman and Trueb's book is a thorough account of anatomy, evolution, and development.

Recently, analyses of genomes have identified tunicates, including sea squirts, as the closest living relatives of vertebrate animals. See F. Delsuc et al., "Tunicates and Not Cephalochordates Are the Closest Living Relatives of Vertebrates," *Nature* 439 (2006): 965–68. Our understanding of vertebrate origins relies also on another living creature, amphioxus, whose genome is discussed in L. Z. Holland et al., "The Amphioxus Genome Illuminates Vertebrate Origins and Cephalochordate Biology," *Genome Research* 18 (2008): 1100–11.

For a general review of Garstang's hypothesis and the problem of vertebrate origins, see Henry Gee, *Across the Bridge: Understanding the Origin of Vertebrates* (Chicago: University of Chicago Press, 2018).

The iconic photo by Naef has generated significant discussion over the years. There is little doubt that he used mounted taxidermy specimens. See, most recently, Richard Dawkins, *The Greatest Show on Earth* (New York: Free Press, 2010). While the postures were likely posed, the similarity of the proportions of the cranial vault, face, and position of the foramen magnum between juvenile chimps and humans has been shown quantitatively in the references below.

The most prominent proponents of human paedomorphosis were Ashley Montagu, *Growing Young* (New York: Greenwood Press, 1989); and Stephen Jay Gould, *Ontogeny and Phylogeny* (Cambridge, MA: Belknap Press, 1985). An opposing view is B. T. Shea, "Heterochrony in Human Evolution: The Case for Neoteny Reconsidered," *Yearbook of Physical Anthropology* 32 (1989): 69–101. While certain traits seem to be paedomorphic, others, such as bipedality, do not.

D'Arcy Wentworth Thompson, *On Growth and Form* (New York: Dover, 1992), originally published in 1917, launched a revolution in quantitative biology. Since his time, the field of morphometrics, the quantitative analysis of shape changes, has been an active area of inquiry.

The importance of the neural crest in development and evolution is reviewed in C. Gans and R. G. Northcutt, "Neural Crest and the Origin of Vertebrates: A New Head," *Science* 220 (1983): 268–73; and Brian Hall, *The Neural Crest in Development and Evolution* (Amsterdam: Springer, 1999).

Julia Platt's work and life is discussed in S. J. Zottoli and E. Seyfarth, "Julia B. Platt (1857–1935): Pioneer Comparative Embryologist and Neuroscientist," *Brain, Behavior and Evolution* 43 (1994): 92–106.

3. MAESTRO IN THE GENOME

The apocryphal quote is taken from J. D. Watson, *The Double Helix* (New York: Touchstone, 2001). Watson and Crick's full quote appeared in a *two-page* paper announcing the finding to science: "We wish to suggest a structure for the salt of deoxyribose nucleic acid (D.N.A.). This structure has novel features which are of considerable biological interest." J. D. Watson and F. Crick, "A Structure for Deoxyribose Nucleic Acid," *Nature* 171 (1953): 737–38.

The story of uncovering the workings of DNA and the ways it makes proteins is discussed in Matthew Cobb, *Life's Greatest Secret: The Race to Crack the Genetic Code* (New York: Basic Books, 2015). See also the classic work by Horace Freeland Judson, *The Eighth Day of Creation: Makers of the Revolution in Biology* (New York: Simon and Schuster, 1979).

Zuckerkandl and Pauling launched their new approach in a series of papers in the mid-1960s. Major ones include E. Zuckerkandl and L. Pauling, "Molecules as Documents of Evolutionary History," *Journal of Theoretical Biology* 8 (1965): 357–66; and E. Zuckerkandl and L. Pauling, "Evolutionary Divergence and Convergence in Proteins," 97–166, in V. Bryson and H. J. Vogel, eds., *Evolving Genes and Proteins* (New York: Academic Press, 1965).

Zuckerkandl and Pauling sought to do more than uncover the relationships between species. They proposed to use the differences in proteins and genes as a kind of clock to tell how long species had been evolving independently of one another. If rates of change in the sequence of a protein are relatively constant over long timescales, then differences in proteins carry a way to interpret time.

The molecular clock hypothesis assumes that over long periods of time, the changes in the sequence of amino acids in a protein will be constant. One way of applying this concept relies on understanding amino acid sequences. Let's take a completely hypothetical example, comparing a species of frog, a monkey, and a human. We would begin by sequencing the proteins. Then we would count the number of amino acids that differ between each of the species. Let's say we're looking at a protein in the skin, and the frog protein differs from both the human and the monkey one by eighty amino acids. Humans and monkeys differ by only thirty. To deploy the molecular clock, we would need to have a fossil date to fix the rate of amino acid change; then we could apply that rate to places where we don't have fossils.

Let's assume we have a fossil that suggests that frogs, monkeys, and people shared a common ancestor 400 million years ago. To calibrate the clock, we would divide 80 by 400 to give a rate of protein change of 0.2 percent over one million years. With this number, we could then calculate how long ago humans and monkeys shared a common ancestor by multiplying 0.2 times 30, to give six million years. This example is hypothetical, but it shows how we would first sequence the proteins, count the amino acid differences among them, use a fossil to estimate the rate of protein change, then apply that rate to understand ages of events for which we do not have fossils.

The account of Zuckerkandl and Pauling's attempt to write an outrageous paper as well as the general historical context of their work is discussed in G. Morgan, "Émile Zuckerkandl, Linus Pauling, and the Molecular Evolutionary Clock," *Journal of the History of Biology* 31 (1998): 155–78. Their resulting paper is E. Zuckerkandl and L. Pauling, "Molecular Disease, Evolution and Genic Heterogeneity," 189–225, in Michael Kasha and Bernard Pullman, eds., *Horizons in Biochemistry: Albert Szent-Györgyi Dedicatory Volume* (New York: Academic Press, 1962).

For an oral history with Émile Zuckerkandl, see "The Molecular Clock," https://authors.library.caltech.edu/5456/1/hrst.mit.edu/hrs/evolution/public/clock/zuckerkandl.html.

Allan Wilson and Mary-Claire King pursued this approach in their work. They were originally following up on an important and controversial molecular clock paper that suggested humans and chimpanzees had a relatively recent common ancestry. That paper is A. Wilson and V. Sarich, "A Molecular Time Scale for Human Evolution," *Proceedings of the National Academy of Sciences* 63 (1969): 1088–93. Their goal was to add more proteins to this analysis to calibrate that clock more precisely. King's epic paper is M. C. King and A. C. Wilson, "Evolution at Two Levels in Humans and Chimpanzees," *Science* 188 (1975): 107–16. The two levels they were referring to were evolution at the level of protein coding and evolution at the level of gene regulation, i.e., the switches. Their data suggested that many of the differences between humans and chimpanzees are due to differences in when and where genes are active; hence, gene regulation.

More recent confirmation of their work is described in Kate Wong, "Tiny Genetic Differences Between Humans and Other Primates Pervade the Genome," *Scientific American*, September 1, 2014; and K. Prüfer et al., "The Bonobo Genome Compared with Chimpanzee and Human Genomes," *Nature* 486 (2012): 527–31.

Several web resources cover the history and impact of the Human Genome Project: "The Human Genome Project (1990–2003)," The Embryo Project Encyclopedia, https://embryo.asu.edu/pages/human-genome-project-1990-2003; "What Is the Human Genome Project?," National Human Genome Research Institute, https://www.genome.gov/12011238/an-overview-of-the-human-genome-project/; and https://www.nature.com/scitable/topicpage/sequencing-human-genome-the-contributions-of-francis-686.

Major scientific papers on the project include International Human Genome Sequencing Consortium, "Finishing the Euchromatic Sequence of the Human Genome," *Nature* 431 (2004): 931–45; and International Human Genome Sequencing Consortium, "Initial Sequencing and Analysis of the Human Genome," *Nature* 409 (2001): 860–921.

Some relevant books on the Human Genome Project include Daniel J. Kevles and Leroy Hood, eds., *The Code of Codes* (Cambridge, MA: Harvard University Press, 2000); and James Shreeve, *The Genome War: How Craig Venter Tried to Capture the Code of Life and Save the World*

(New York: Random House, 2004). A firsthand account is John Craig Venter, *A Life Decoded: My Genome: My Life* (New York: Viking Press, 2007).

The structure of the genome and the number of genes have a large literature, including a number of prominent multi-investigator projects. An introductory sampling, with good bibliographies, includes A. Prachumwat and W.-H. Li, "Gene Number Expansion and Contraction in Vertebrate Genomes with Respect to Invertebrate Genomes," *Genome Research* 18 (2008): 221–32; and R. R. Copley, "The Animal in the Genome: Comparative Genomics and Evolution," *Philosophical Transactions of the Royal Society, B* 363 (2008): 1453–61. The journal *Nature* has a good introductory website: https://www.nature.com/scitable/topicpage/eukaryotic-genome-complexity-437.

Powerful genome browsers allow scientists to compare genes and genomes of different species. Some of the most frequently used include ENSEMBL https://useast.ensembl.org/; VISTA, http://pipeline.lbl.gov/cgi-bin/gateway2; and the BLAST search tool, https://blast.ncbi.nlm.nih.gov/Blast.cgi. Check them out. They place a world of discovery at your fingertips.

François Jacob and Jacques Monod's classic is one of the greatest papers in biology: "Genetic Regulatory Mechanisms in the Synthesis of Proteins," *Journal of Molecular Biology* 3 (1961): 318–56. It is challenging for the novice to read. For a thorough yet readable breakdown, see this classic in scientific communication: Horace Freeland Judson, *The Eighth Day of Creation: Makers of the Revolution in Biology* (New York: Simon and Schuster, 1979).

For the incredible backdrop of Jacob and Monod's work, see the gripping and authoritative account by Sean B. Carroll, *Brave Genius: A Scientist, a Philosopher, and Their Daring Adventures from the French Resistance to the Nobel Prize* (New York: Norton, 2013). I thought I knew everything about them, but this book opened an entire world for me.

Sean B. Carroll also wrote the classic on how gene regulation can impact evolution: *Endless Forms Most Beautiful: The New Science of Evo Devo* (New York: Norton, 2006).

The role of *Sonic hedgehog* in limb anomalies is discussed in E. Anderson et al., "Human Limb Abnormalities Caused by Disruption of Hedgehog Signaling," *Trends in Genetics* 28 (2012): 364–73. Anomalies come about by changing the activity of *Sonic* or by disrupting the pathway of genes with which *Sonic* interacts.

The work on the long-range switch, more formally known as a long-range enhancer, is in a series of beautiful papers: L. A. Lettice et al., "The Conserved *Sonic hedgehog* Limb Enhancer Consists of Discrete Functional Elements That Regulate Precise Spatial Expression," *Cell Reports* 20 (2017): 1396–408; L. A. Lettice et al., "A Long-Range *Shh* Enhancer Regulates Expression in the Developing Limb and Fin and Is Associated with Preaxial Polydactyly," *Human Molecular Genetics* 12 (2003): 1725–35; and R. Hill and L. A. Lettice, "Alterations to the Remote Control of *Shh* Gene Expression Cause Congenital Abnormalities," *Philosophical Transactions of the Royal Society, B* 368 (2013), http://doi.org/10.1098/rstb.2012.0357.

Many of these long-range switches are now known. On their general biology and impacts on development and evolution, see A. Visel et al., "Genomic Views of Distant-Acting Enhancers," *Nature* 461 (2009): 199–205; H. Chen et al., "Dynamic Interplay Between Enhancer-Promoter Topology and Gene Activity," *Nature Genetics* 50 (2018): 1296–303; and A. Tsai and J. Crocker, "Visualizing Long-Range Enhancer-Promoter Interaction," *Nature Genetics* 50 (2018): 1205–6.

The reduction in snake limbs and the correlation to changes in the *Sonic* long-range enhancer is discussed in E. Z. Kvon et al., "Progressive Loss of Function in a Limb Enhancer During Snake Evolution," *Cell* 167 (2016): 633–42.

The role of changes in genetic regulatory elements (switches) has a large literature. See M. Rebeiz and M. Tsiantis, "Enhancer Evolution and the Origins of Morphological Novelty," *Current Opinion in Genetics and Development* 45 (2017): 115–23; and Sean B. Carroll, *Endless Forms Most Beautiful: The New Science of Evo Devo* (New York: Norton, 2006). For the stickleback example, see Y. F. Chan et al., "Adaptive Evolution of Pelvic Reduction in Sticklebacks by Recurrent Deletion of a *Pitx1* Enhancer," *Science* 327 (2010): 302–5.

4. BEAUTIFUL MONSTERS

Thomas Soemmerring was a polymath who described one of the first flying reptiles, pterosaurs, designed telescopes, developed vaccines, and analyzed mutants. His classic work on developmental anomalies is S. T. von Soemmerring, *Abbildungen und Beschreibungen einiger Misgeburten die sich ehemals auf dem anatomischen Theater zu Cassel befanden* (Mainz: kurfürstl. privilegirte Universitätsbuchhandlung, 1791).

An influential paper on how monsters—developmental anomalies—can be deeply informative is P. Alberch, "The Logic of Monsters: Evidence for Internal Constraint in Development and Evolution," *Geobios* 22 (1989): 21–57.

For classical interpretations of developmental anomalies and teratology, see Dudley Wilson, *Signs and Portents: Monstrous Births from the Middle Ages to the Enlightenment* (New York: Routledge, 1993).

On the enduring contribution of Geoffroy and Isidore Saint-Hilaire to understanding developmental anomalies, see A. Morin, "Teratology from Geoffroy Saint Hilaire to the Present," *Bulletin de l'Association des anatomistes (Nancy)* 80 (1996): 17–31 (in French).

For an informative website on the history and impact of studies of teratology on biology and medicine, see "A New Era: The Birth of a Modern Definition of Teratology in the Early 19th Century," New York Academy of Medicine, https://nyam.org/library/collections-and

-resources/digital-collections-exhibits/digital-telling-wonders/new-era-birth-modern-definition-teratology-early-19th-century/.

William Bateson's classic work on variation is *Materials for the Study of Variation Treated with Especial Regard to Discontinuity in the Origin of Species* (London: Macmillan, 1894).

One of T. H. Morgan's former students, an eminence in his own right, wrote his National Academy of Sciences Biographical Memoir: A. H. Sturtevant, *Thomas Hunt Morgan, 1866–1945: A Biographical Memoir* (Washington, DC: National Academy of Sciences, 1959), available online at http://www.nasonline.org/publications/biographical-memoirs/memoir-pdfs/morgan-thomas-hunt.pdf.

Calvin Bridges was the subject of a 2014 biopic, *The Fly Room*, reviewed in Ewen Callaway, "Genetics: Genius on the Fly," *Nature* 516 (December 11, 2014), online at https://www.nature.com/articles/516169a.

Cold Spring Harbor Laboratory maintains a biographical website devoted to Calvin Bridges: Calvin Blackman Bridges, Unconventional Geneticist (1889–1938), at http://library.cshl.edu/exhibits/bridges.

For a history of Lewis and Bridges's work, see I. Duncan and G. Montgomery, "E. B. Lewis and the Bithorax Complex," pts. 1 and 2, *Genetics* 160 (2002): 1265–72, and 161 (2002): 1–10. Lewis was initially more interested in gene duplications than in development; hence his interest in this region of the chromosome.

The banding patterns on chromosomes as a road map to *Bithorax* and other mutants is described in C. B. Bridges, "Salivary Chromosome Maps: With a Key to the Banding of the Chromosomes of *Drosophila melanogaster*," *Journal of Heredity* 26 (1935): 60–64; and C. B. Bridges and T. H. Morgan, *The Third-Chromosome Group of Mutant Characters of Drosophila melanogaster* (Washington, DC: Carnegie Institution, 1923).

Edward Lewis's classic paper is E. B. Lewis, "A Gene Complex Controlling Segmentation in Drosophila," *Nature* 276 (1978): 565–70.

The homeobox discovery was made in parallel by W. McGinnis et al.,

"A Conserved DNA Sequence in Homoeotic Genes of the *Drosophila* Antennapedia and Bithorax Complexes," *Nature* 308 (1984): 428–33; and by M. Scott and A. Weiner, "Structural Relationships Among Genes That Control Development: Sequence Homology Between the Antennapedia, Ultrabithorax, and Fushi Tarazu Loci of Drosophila," *Proceedings of the National Academy of Sciences* 81 (1984): 4115–19.

The homeobox discovery and its implications for evolution is given a full account, with references, in Sean B. Carroll, *Endless Forms Most Beautiful: The New Science of Evo Devo* (New York: Norton, 2006). Ed Lewis retrospectively reviewed the problem in E. B. Lewis, "Homeosis: The First 100 Years," *Trends in Genetics* 10 (1994): 341–43.

Patel's work with *Parhyale* is described in A. Martin et al., "CRISPR/Cas9 Mutagenesis Reveals Versatile Roles of *Hox* Genes in Crustacean Limb Specification and Evolution," *Current Biology* 26 (2016): 14–26; and J. Serano et al., "Comprehensive Analysis of *Hox* Gene Expression in the Amphipod Crustacean *Parhyale hawaiensis*," *Developmental Biology* 409 (2016): 297–309.

On the role of the homeobox genes in the development of vertebrae, see D. Wellik and M. Capecchi, "*Hox10* and *Hox11* Genes Are Required to Globally Pattern the Mammalian Skeleton," *Science* 301 (2003): 363–67; and D. Wellik, "*Hox* Patterning of the Vertebrate Axial Skeleton," *Developmental Dynamics* 236 (2007): 2454–63.

The "hand genes" are known more precisely as *Hoxa-13* and *Hoxd-13*. The paper describing their deletion in mice is C. Fromental-Ramain et al., "*Hoxa-13* and *Hoxd-13* Play a Crucial Role in the Patterning of the Limb Autopod," *Development* 122 (1996): 2997–3011.

Tetsuya Nakamura and Andrew Gehrke's studies of homeobox genes in fin development are contained in T. Nakamura et al., "Digits and Fin Rays Share Common Developmental Histories," *Nature* 537 (2016): 225–28. Their work was also reported in Carl Zimmer, "From Fins into Hands: Scientists Discover a Deep Evolutionary Link," *New York Times*, August 17, 2016.

5. COPYCATS

Vicq d'Azyr is an underappreciated figure in the history of anatomy. He made many of the same observations that Richard Owen did about the similarity of form (such as homology) but never generalized them, so he is not as widely credited with their origin. See R. Mandressi, "The Past, Education and Science. Félix Vicq d'Azyr and the History of Medicine in the 18th Century," *Medicina nei secoli* 20 (2008): 183–212 (in French); and R. S. Tubbs et al., "Félix Vicq d'Azyr (1746–1794): Early Founder of Neuroanatomy and Royal French Physician," *Child's Nervous System* 27 (2011): 1031–34.

A more modern take on this notion of duplicate organs in the body, known as serial homology, is in Günter Wagner, *Homology, Genes, and Evolutionary Innovation* (Princeton, NJ: Princeton University Press, 2018).

The small eyes mutant was first described in Sabra Colby Tice, *A New Sex-linked Character in Drosophila* (New York: Zoological Laboratory, Columbia University, 1913).

Bridges's use of his chromosomal maps to reveal gene duplications is found in "Calvin Bridges, "Salivary Chromosome Maps: With a Key to the Banding of the Chromosomes of *Drosophila melanogaster*," *Journal of Heredity* 26 (1935): 60–64.

The life of Susumu Ohno is covered in U. Wolf, "Susumu Ohno," *Cytogenetics and Cell Genetics* 80 (1998): 8–11; and in Ernest Beutler, "Susumu Ohno, 1928–2000" *Biographical Memoirs* 81 (2012), from the National Academy of Sciences, online at https://www.nap.edu/read/10470/chapter/14.

Ohno's work is in a number of papers and a book that synthesizes his work on duplications: Susumu Ohno, "So Much 'Junk' DNA in Our Genome," 336–70, in H. H. Smith, ed., *Evolution of Genetic Systems* (New York: Gordon and Breach, 1972); Susumu Ohno, "Gene

Duplication and the Uniqueness of Vertebrate Genomes Circa 1970–1999," *Seminars in Cell and Developmental Biology* 10 (1999): 517–22; and Susumu Ohno, *Evolution by Gene Duplication* (Amsterdam: Springer, 1970).

Yves Van de Peer, Eshchar Mizrachi, and Kathleen Marchal, "The Evolutionary Significance of Polyploidy," *Nature Reviews Genetics* 18 (2017): 411–24; and S. A. Rensing, "Gene Duplication as a Driver of Plant Morphogenetic Evolution," *Current Opinion in Plant Biology* 17 (2014): 43–48.

T. Ohta, "Evolution of Gene Families," *Gene* 259 (2000): 45–52; J. Thornton and R. DeSalle, "Gene Family Evolution and Homology: Genomics Meets Phylogenetics," *Annual Reviews of Genomics and Human Genetics* 1 (2000): 41–73; and J. Spring, "Genome Duplication Strikes Back," *Nature Genetics* 31 (2002): 128–29.

There are many examples of gene families and their evolution. One from the opsin genes used in seeing is a nice example. See R. M. Harris and H. A. Hoffman, "Seeing Is Believing: Dynamic Evolution of Gene Families," *Proceedings of the National Academy of Sciences* 112 (2015): 1252–53.

Homeobox genes are another case of a gene family that arose by gene duplication. For different perspectives on the mechanisms and the impact of this duplication, see P. W. H. Holland, "Did Homeobox Gene Duplications Contribute to the Cambrian Explosion?," *Zoological Letters* 1 (2015): 1–8; G. P. Wagner et al., "*Hox* Cluster Duplications and the Opportunity for Evolutionary Novelties," *Proceedings of the National Academy of Sciences* 100 (2003): 14603–6; and N. Soshnikova et al., "Duplications of *Hox* Gene Clusters and the Emergence of Vertebrates," *Developmental Biology* 378 (2013): 194–99.

Notch signaling and the duplication of genes in brain evolution was the subject of two papers published independently: I. T. Fiddes et al., "Human-Specific *NOTCH2NL* Genes Affect Notch Signaling and Cortical Neurogenesis," *Cell* 173 (2018): 1356–69; and I. K. Suzuki et al., "Human-Specific *NOTCH2NL* Genes Expand Cortical Neurogenesis Through Delta/Notch Regulation," *Cell* 173 (2018): 1370–84.

Roy Britten's life is recounted by his longtime collaborator in Eric Davidson, "Roy J. Britten, 1919–2012: Our Early Years at Caltech," *Proceedings of the National Academy of Sciences* 109 (2012): 6358–59. Davidson and Britten together published a speculative paper on the meaning of these sequences that was well ahead of its time and spawned research by a generation of scientists: R. J. Britten and E. H. Davidson, "Repetitive and Non-Repetitive DNA Sequences and a Speculation on the Origins of Evolutionary Novelty," *Quarterly Review of Biology* 46 (1971): 111–38.

Britten's paper describing the repeats and the techniques he used to find them is R. J. Britten and D. E. Kohne, "Repeated Sequences in DNA," *Science* 161 (1968): 529–40. A simpler translation of the work and its context is R. Andrew Cameron, "On DNA Hybridization and Modern Genomics," at https://onlinelibrary.wiley.com/doi/pdf/10.1002/mrd.22034.

Manyuan Long's lab group described its work on the origin of new genes in W. Zhang et al., "New Genes Drive the Evolution of Gene Interaction Networks in the Human and Mouse Genomes," *Genome Biology* 16 (2015): 202–26. The origin of new genes is an active area of inquiry. While many new genes arise by gene duplication, some do not, and the mechanisms for this are still under active inquiry. For an exemplar with references, see L. Zhao et al., "Origin and Spread of De Novo Genes in *Drosophila melanogaster* Populations," *Science* 343 (2014): 769–72.

McClintock's jumping gene discovery is first described in Barbara McClintock, "The Origin and Behavior of Mutable Loci in Maize," *Proceedings of the National Academy of Sciences* 36 (1950): 344–55. For a retrospective celebration and explanation of the paper, see S. Ravindran, "Barbara McClintock and the Discovery of Jumping Genes," *Proceedings of the National Academy of Sciences* 109 (2012): 20198–99.

On the discovery and workings of jumping genes, see L. Pray and K. Zhaurova, "Barbara McClintock and the Discovery of Jumping Genes (Transposons)," *Nature Education* 1 (2008): 169.

The National Library of Medicine has an online repository of McClintock's papers, including her quotes used here and the quote by Nixon at her National Medal of Science ceremony: https://profiles.nlm.nih.gov/ps/retrieve/Narrative/LL/p-nid/52.

6. OUR INNER BATTLEFIELD

Ernst Mayr's classic book is *Animal Species and Evolution* (Cambridge, MA: Harvard University Press, 1963).

Richard Goldschmidt's book is *The Material Basis of Evolution* (New Haven, CT: Yale University Press, 1940). The paper that so enraged Mayr is Goldschmidt, "Evolution as Viewed by One Geneticist," *American Scientist* 40 (1952): 84–98.

For Goldschmidt's life, see Curt Stern, *Richard Benedict Goldschmidt, 1878–1958: A Biographical Memoir* (Washington, DC: National Academy of Sciences, 1967), at http://www.nasonline.org/publications/biographical-memoirs/memoir-pdfs/goldschmidt-richard.pdf.

The era when Mayr did his major work is known as the time of the Evolutionary Synthesis; it culminated in the late 1940s, when findings from genetics were incorporated into the fields of taxonomy, paleontology, and comparative anatomy. During our continued teas, Mayr often spoke of a whole new synthesis being on the horizon in the 1990s, one that would extend the work of his generation into molecular biology and developmental genetics. Accordingly, he encouraged the graduate students in his retinue to stay current in that scientific literature.

Ronald Fisher's enormously influential work was *The Genetical Theory of Natural Selection* (London: Clarendon Press, 1930).

Vincent Lynch's papers are V. J. Lynch et al., "Ancient Transposable Elements Transformed the Uterine Regulatory Landscape and Transcriptome During the Evolution of Mammalian Pregnancy," *Cell Reports* 10 (2015): 551–61; and V. J. Lynch et al., "Transposon-Mediated Rewiring of Gene Regulatory Networks Contributed to the Evolution of Pregnancy in Mammals," *Nature Genetics* 43 (2011): 1154–58.

Lynch reviewed the general problem in G. P. Wagner and V. J. Lynch, "The Gene Regulatory Logic of Transcription Factor Evolution," *Trends in Ecology and Evolution* 23 (2008): 377–85; and G. P. Wagner and V. J. Lynch, "Evolutionary Novelties," *Current Biology* 20 (2010): 48–52. The inspiration for this work was McClintock herself in B. McClintock, "The Origin and Behavior of Mutable Loci in Maize," *Proceedings of the National Academy of Sciences* 36 (1950): 344–55; and the seminal paper by R. J. Britten and E. H. Davidson, "Repetitive and Non-Repetitive DNA Sequences and a Speculation on the Origins of Evolutionary Novelty," *Quarterly Review of Biology* 46 (1971): 111–38.

The conversion of jumping genes into useful parts of the genome (their so-called domestication) is an active area of research. A sampling of papers and references includes D. Jangam et al., "Transposable Element Domestication as an Adaptation to Evolutionary Conflicts," *Trends in Genetics* 33 (2017): 817–31; and E. B. Chuong et al., "Regulatory Activities of Transposable Elements: From Conflicts to Benefits," *Nature Reviews Genetics* 18 (2017): 71–86.

A good review of the syncytin work is C. Lavialle et al., "Paleovirology of 'Syncytins,' Retroviral env Genes Exapted for a Role in Placentation," *Philosophical Transactions of the Royal Society of London, B* 368 (2013): 20120507; and H. S. Malik, "Retroviruses Push the Envelope for Mammalian Placentation," *Proceedings of the National Academy of Sciences* 109 (2012): 2184–85. The syncytin discoveries are in S. Mi et al., "Syncytin Is a Captive Retroviral Envelope Protein Involved in Human Placental Morphogenesis" *Nature* 403 (2000): 785–89; J. Denner, "Expression and Function of Endogenous Retroviruses in the Placenta," *APMIS* 124 (2016): 31–43; A. Dupressoir et al., "Syncytin-A Knockout Mice Demonstrate the Critical Role in Placentation of a Fusogenic, Endogenous Retrovirus-Derived, Envelope Gene," *Proceedings of the National Academy of Sciences* 106 (2009): 12127–32; and A. Dupressoir et al., "A Pair of Co-Opted Retroviral Envelope Syncytin Genes Is Required for Formation of the Two-Layered Murine Placental Syncytiotrophoblast," *Proceedings of the National Academy of Sciences* 108 (2011): 1164–73.

For a general review of the role of retroviruses in the evolution of the placenta, see D. Haig, "Retroviruses and the Placenta," *Current Biology* 22 (2012): 609–13.

Syncytins have also now been found in other species that have placenta-like structures, such as lizards. See G. Cornelis et al., "An Endogenous Retroviral Envelope Syncytin and Its Cognate Receptor Identified in the Viviparous Placental *Mabuya* Lizard," *Proceedings of the National Academy of Sciences* 114 (2017): E10991–E11000.

The search for long-dead or domesticated viruses is a field unto itself, known as paleovirology. For more information, see M. R. Patel et al., "Paleovirology—Ghosts and Gifts of Viruses Past," *Current Opinion in Virology* 1 (2011): 304–9; and J. A. Frank and C. Feschotte, "Co-option of Endogenous Viral Sequences for Host Cell Function," *Current Opinion in Virology* 25 (2017): 81–89.

Jason Shepherd's work with *Arc* is in E. D. Pastuzyn et al., "The Neuronal Gene *Arc* Encodes a Repurposed Retrotransposon Gag Protein That Mediates Intercellular RNA Transfer," *Cell* 172 (2018): 275–88. Ed Yong reviewed the paper for a more general audience in "Brain Cells Share Information with Virus-Like Capsules," *Atlantic* (January 2018).

7. LOADED DICE

The book that emerged from Gould's lectures was Stephen Jay Gould, *Wonderful Life: The Burgess Shale and the Nature of History* (New York: Norton, 1989).

For Ray Lankester's work on degeneration and multiples in evolution, see E. R. Lankester, *Degeneration: A Chapter in Darwinism* (London: Macmillan, 1880); and E. R. Lankester, "On the Use of the Term 'Homology' in Modern Zoology, and the Distinction Between Homogenetic and Homoplastic Agreements," *Annals and Magazine of Natural History* 6 (1870): 34–43.

For a discussion of convergent and parallel evolution, see Simon Conway Morris, *Life's Solution: Inevitable Humans in a Lonely Universe* (Cambridge, UK: Cambridge University Press, 2003). Conway Morris takes the hard stand that all of evolution is inevitable. By contrast, Jonathan Losos, *Improbable Destinies: Fate, Chance and the Future of Evolution* (New York: Riverhead, 2017), is a finely balanced view of the relationship between chance and inevitability.

Good footage of salamander tongue flipping is at https://www.youtube.com/watch?v=mRrIITcUeBM.

A scientific breakdown of the anatomy behind this amazing feature is S. M. Deban et al., "Extremely High-Power Tongue Projection in Plethodontid Salamanders," *Journal of Experimental Biology* 210 (2007): 655–67.

Wake's original paper on tongue projection is a classic: R. E. Lombard and D. B. Wake, "Tongue Evolution in the Lungless Salamanders, Family Plethodontidae IV. Phylogeny of Plethodontid Salamanders and the Evolution of Feeding Dynamics," *Systematic Zoology* 35 (1986): 532–51.

The remarkable multiple evolution of tongue projection is shown in D. B. Wake et al., "Transitions to Feeding on Land by Salamanders Feature Repetitive Convergent Evolution," 395–405, in K. Dial, N. Shubin, and E. L. Brainerd, eds., *Great Transformations in Vertebrate Evolution* (Chicago: University of Chicago Press, 2015).

The frozen salamander analysis is in N. H. Shubin et al., "Morphological Variation in the Limbs of *Taricha Granulosa* (Caudata: Salamandridae): Evolutionary and Phylogenetic Implications," *Evolution* 49 (1995): 874–84. The evolutionary interpretation and predictability of their patterns is discussed in N. Shubin and D. B. Wake, "Morphological Variation, Development, and Evolution of the Limb Skeleton of Salamanders," 1782–808, in H. Heatwole, ed., *Amphibian Biology* (Sydney: Surrey Beatty, 2003); N. Shubin and P. Alberch, "A Morphogenetic Approach to the Origin and Basic Organization of the Tetrapod Limb," *Evolutionary Biology* 20 (1986): 319–87; N. B. Fröbisch and N. Shubin, "Salamander Limb Development: Integrating Genes, Morphology, and Fossils," *Developmental Dynamics* 240 (2011): 1087–99; N. Shubin and D. Wake, "Phylogeny, Variation and Morphological Integration," *American Zoologist* 36 (1996): 51–60; and N. Shubin, "The Origin of Evolutionary Novelty: Examples from Limbs," *Journal of Morphology* 252 (2002): 15–28.

Wake wrote some general papers on how multiples in evolution reveal general mechanisms of change: D. B. Wake et al., "Homoplasy: From Detecting Pattern to Determining Process and Mechanism of Evolution," *Science* 331 (2011): 1032–35; and D. B. Wake, "Homoplasy: The Result of Natural Selection, or Evidence of Design Limitations?," *American Naturalist* 138 (1991): 543–61.

Another scholarly review of multiples in evolution is B. K. Hall, "Descent with Modification: The Unity Underlying Homology and Homoplasy as Seen Through an Analysis of Development and Evolution," *Biological Reviews of the Cambridge Philosophical Society* 78 (2003): 409–33.

The work on Caribbean lizards is reviewed in Jonathan Losos,

Improbable Destinies: Fate, Chance and the Future of Evolution (New York: Riverhead, 2017).

Rich Lenski's laboratory at Michigan State University has been carrying out a long-term experiment with bacteria that began in 1998. This venture, bold at the time, has allowed for direct observation of many major kinds of evolutionary change, giving us the tools to see these events in action. This review reveals the complex relationship of determinism and contingency in evolution: Z. Blount, R. Lenski, and J. Losos, "Contingency and Determinism in Evolution: Replaying Life's Tape," *Science* 362:6415 (2018): doi: 10.1126/scienceaam5979.

8. MERGERS AND ACQUISITIONS

Lynn Margulis's original paper is L. [Margulis] Sagan, "On the Origin of Mitosing Cells," *Journal of Theoretical Biology* 14 (1967): 225–74. Her wide-ranging book on her theory is Lynn Margulis, *Symbiosis in Cell Evolution: Life and Its Environment on the Early Earth* (San Francisco: Freeman, 1981). Her retrospective quote is taken from a 2011 interview in *Discover* magazine, available online at http://discovermagazine.com/2011/apr/16-interview-lynn-margulis-not-controversial-right.

For recent perspectives including references, see J. Archibald, *One Plus One Equals One: Symbiosis and the Evolution of Complex Life* (Oxford: Oxford University Press, 2014); L. Eme et al., "Archaea and the Origin of Eukaryotes," *Nature Reviews Microbiology* 15 (2017): 711–23; J. M. Archibald, "Endosymbiosis and Eukaryotic Cell Evolution," *Current Biology* 25 (2015): 911–21; and M. O'Malley, "Endosymbiosis and Its Implications for Evolutionary Theory," *Proceedings of the National Academy of Sciences* 112 (2015): 10270–77.

Compelling and informative resources on the earliest phases of life's history include Andrew Knoll, *Life on a Young Planet: The First Three Billion Years of Evolution on Earth* (Princeton, NJ: Princeton University Press, 2004); Nick Lane, *The Vital Question: Energy, Evolution, and the Origins of Complex Life* (New York: Norton, 2015); and J. William Schopf, *Cradle of Life: The Discovery of Earth's Earliest Fossils* (Princeton, NJ: Princeton University Press, 1999).

Schopf's collaborative work on the carbon isotopic analysis of the Apex Chert structures is in J. W. Schopf et al., "SIMS Analyses of the Oldest Known Assemblage of Microfossils Document Their Taxon-Correlated Carbon Isotope Compositions," *Proceedings of the National Academy of Sciences* 115 (2018): 53–58.

The meaning and evolution of individuality is discussed in a little book that had a big impact: Leo Buss, *The Evolution of Individuality* (Princeton, NJ: Princeton University Press, 1988). Buss focuses on what an individual is and shows how natural selection operates as new individuals and levels of selection emerge.

An approach to the origin of new types of individuals, and their impact on evolution, is in John Maynard-Smith and Eörs Szathmáry, *The Major Transitions in Evolution* (Oxford: Oxford University Press, 1998).

Nicole King's wonderful lecture "Choanoflagellates and the Origin of Animal Multicellularity" is online at https://www.ibiology.org/ecology/choanoflagellates/.

For work on choanoflagellates, see T. Brunet and N. King, "The Origin of Animal Multicellularity and Cell Differentiation," *Developmental Cell* 43 (2017): 124–40; S. R. Fairclough et al., "Multicellular Development in a Choanoflagellate," *Current Biology* 20 (2010): 875–76; R. A. Alegado and N. King, "Bacterial Influences on Animal Origins," *Cold Spring Harbor Perspectives in Biology* 6 (2014): 6:a016162; and D. J. Richter and N. King, "The Genomic and Cellular Foundations of Animal Origins," *Annual Review of Genetics* 47 (2013): 509–37.

A good primer on CRISPR-Cas genome editing, including its history, was co-written by one of its pioneers: Jennifer Doudna and Samuel Sternberg, *A Crack in Creation: Gene Editing and the Unthink-*

able Power to Control Evolution (New York: Houghton Mifflin Harcourt, 2017).

EPILOGUE

Mount Ritchie lies in Victoria Land in Antarctica. We were there as part of a U.S. Antarctic Program project funded by the National Science Foundation Grant 1543367.

ACKNOWLEDGMENTS

This book is dedicated to my late parents, Seymour and Gloria Shubin, for fostering a love of the natural world, a curiosity about how it works, and the importance of telling a good story. For my previous work, my father, a fiction writer who did not find science easily digestible, served as my target audience. If he enjoyed the narrative and appreciated the science, then I knew I was doing things right. His presence remains on every page here.

This is the third book I have done with Kalliopi Monoyios as illustrator. She brings a passion for science and a keen eye for visual storytelling; this book was no exception. She read drafts, pursued permissions, and was invaluable in finding chinks in my storytelling and science. Kapi is online at www.kalliopimonoyios.com and Instagram at kalliopi.monoyios.

Several people generously shared stories of their science, personal history, or ideas. These include Cedric Feschotte, Bob Hill, Mary-Claire King, Nicole King, Chris Lowe, Vinny Lynch, Nipam Patel, Jason Shepherd, and David Wake. John Novembre, Michele Seidl, and Kalliopi Monoyios read portions or drafts and offered important comments. Any misinterpretation of personal stories or errors in the science are, of course, my own.

Members of my laboratory endured absences from the lab for the past three years. I'm grateful to current and past lab members: Noritaka Adachi, Melvin Bonilla, Andrew Gehrke, Katie Mika, Mirna Marinic, Tesuya Nakamura, Atreyo Pal, Joyce Pieretti, Igor Schneider, Gayani

Senevirathne, Tom Stewart, and Julius Tabin for pushing and inspiring me by their own examples to do ever better science. I am fortunate to have scientific collaborators who catalyze both my science and the ways I communicate it. These include members of my recent polar field teams as well as those who've collaborated with me or coached me on molecular biology: Sean Carroll, Ted Daeschler, Marcus Davis, John Long, Adam Maloof, Tim Senden, José-Luis Gomez Skarmeta, and Cliff Tabin.

Nothing ever begins when you think it does. In one way or another, these ideas have been in my mind since my years in graduate school at Harvard, and later at Berkeley, when I had the chance to interact with people whose ideas and approaches profoundly affected my worldview. These include Pere Alberch, Stephen Jay Gould, Ernst Mayr, and David Wake. Fellow graduate students from those days had an enormous impact on me, including Annie Burke, Edwin Gilland, and Greg Mayer. My thinking was crystallized by discussions and collegial debates with all of these individuals.

Much of this book was written while I was serving on the leadership of the Marine Biological Laboratory at Woods Hole, Massachusetts (MBL). The MBL is a special place to learn about and do science, drawing a remarkable community of resident and visiting scientists in the life sciences each year. Writing chapters of this book in the MBL's Lillie Library connected me to previous denizens who formed the basis for several chapters: Julia Platt, O. C. Whitman, T. H.Morgan, and Émile Zuckerkandl. The Wellfleet, Eastham, Orleans, and Truro libraries served as quiet and refreshing venues in which to write each summer.

My agents, Katinka Matson, Max Brockman, and Russell Weinberger, were a continual source of support, shepherding this project along. Dan Frank has edited three books of mine, and each one has been a master class in learning the art of writing and publication. Dan encouraged me, prodded me to improve, and was patient with me throughout the process. My British editor, Sam Carter, has been a wonderful source of encouragement. Dan Frank's assistant, Vanessa Rae Haughton, cheerfully guided the project from manuscript to book. The remarkable production and copyediting teams at Pantheon—Roméo Enriquez, Ellen Feldman, Janet Biehl, Chuck Thompson, and Laura Starrett—were nothing short of heroic in their work. Thanks to Anna Knighton for her text design, and to Perry De La Vega, who turned the book's themes into a wonderful cover. Michiko Clark and the publicity team at Pantheon have been a joy to work with.

My family has lived with this project for almost five years, enduring absences and endless discussions of fossils, DNA, and the history of life. My wife, Michele Seidl, and children, Nathaniel and Hannah, were always by my side on a path that was much like evolution itself: full of twists, turns, surprises, and, of course, wonder.

ILLUSTRATION CREDITS

All figures unless otherwise noted are in the public domain.

12 © Kalliopi Monoyios
14 © The Metropolitan Museum of Art; used with permission
23 *Deinonychus* © Paul Heaston, used with permission; silhouette © Kalliopi Monoyios
25 From F. M. Smithwick, R. Nicholls, I. C. Cuthill, and J. Vinther (2017), "Countershading and Stripes in the Theropod Dinosaur Sinosauropteryx Reveal Heterogeneous Habitats in the Early Cretaceous Jehol Biota" (http://www.cell.com/currentbiology/fulltext/S0960-9822(17)31197-1), *Current Biology*. DOI:10.1016/j.cub.2017.09.032 (https://doi.org/10.1016/j.cub.2017.09.032), used under CC BY 4.0 International
40 Scott Polar Research Institute, University of Cambridge; used with permission
42 From Walter Garstang, *Larval Forms and Other Verses*
45 © Kalliopi Monoyios
48 © Kalliopi Monoyios
57 Courtesy of Stanford University's Hopkins Marine Station
79 © Kalliopi Monoyios
80 By Marc Averette, used under CC BY 3.0 Unported
85 © Kalliopi Monoyios
99 © Kalliopi Monoyios

100 © 2011 Wolfgang F. Wülker, from W. F. Wülker et al., "Karyotypes of *Chironomus* Meigen (Diptera: Chironomidae) Species from Africa," *Comparative Cytogenetics* 5(1): 23–46, https://doi.org/10.3897/compcytogen.v5i1.975, used under CC BY 3.0
101 From the archives of the Smithsonian Institution; used with permission
102 © Kalliopi Monoyios
104 Image by Howard Lipschitz, reprinted with kind permission of Springer Nature, The Netherlands
107 © Kalliopi Monoyios
110 © Kalliopi Monoyios
115 © Kalliopi Monoyios
118 © Kalliopi Monoyios
122 Andrew Gehrke and Tetsuya Nakamura, University of Chicago
128 Courtesy of City of Hope Archives; used with permission
141 Courtesy of Barbara McClintock Collection at Cold Spring Harbor Laboratory; used with permission
152 Courtesy of Vincent J. Lynch; used with permission
171 Color lithograph after Sir L. Ward, Wellcome Library, London, used under CC BY 4.0
174 Courtesy of David Wake; used with permission
179 © Kalliopi Monoyios
187 Gayani Senevirathne, University of Chicago; used with permission
189 © Kalliopi Monoyios
194 Courtesy of Boston University Photography; used with permission
196 © Kalliopi Monoyios
205 Courtesy of Nicole King; used with permission

INDEX

Page numbers in *italic* refer to illustrations.

Allen, Joel Asaph, 183
Allen's Rule, 183
"Almost All Human Genes Arose by Duplication" (Britten), 140
ALU sequence, 140, 144–45
Alzheimer's disease, 162
Ambystoma, 30–31
amino acids, 62–69
ammocoete, 42, 46
amphipods, 111–16
Animals, Species and Evolution (Mayr), 146–47
annular tendon of Zinn, 124
Antennapedia mutant, *107*, 107–8
Apex Chert formation, 198–200
Arc gene, 162–66
Archaeopteryx, 17–19, 25–26, 41
Aristotle, 32–33, 46
Augustine, 28–29
"The Axolotl and the Ammocoete" (Garstang), 42, 46
Bachmann's bundle, 124
bacteria
  genome of, 77–80, 195
  incorporation into microbes of, 193–97, 201–2
  jumping genes in, 144
  resistance to viruses of, 211–12
  in salty habitats, 210–13
bases, 60–61
Bateson, William, 94–95, 118–19
bats, 182
Bergmann, Carl, 183
Bergmann's Rule, 183
bichirs, 11, 15
biological multiples, 173–90
  See also contingency and chance
birds. *See* embryonic development; flight
Bithorax fruit flies, *102*, 102–6, 110–11, 135

Blair, Tony, 75
Blake, William, 218
blue-green algae, 195–99, 201
bodies
  building of, 112–16, 202–7
  organization of, 207–9
Bolitoglossa, 177
bones
  of dinosaurs, 22–24, 26
  of humans, 150
  specialization for flight of, 25–26, 182
Bonnet, Charles, 32–33
brain size, 136–38
Bridges, Calvin, 100–104, *101*, 111, 119, 126–27, 148
Britten, Roy, 139–40, 144

cadherins, 204, 206
Caenorhabditis elegans, 76, 202
cancer, 208–9
Capra, Frank, 169–70
carbon content, 199–200
cartilage, 150
Cas9 (molecular scalpel), 212–13
Castle, William E., 97–98
cataclysm, 168–70
catfish, 15
Cerotodus, 13
Champollion, Jean-François, 10
Cherry-Garrard, Apsley, *40*, 40–41
cherts, 198–201
chicken embryos, 32–35, 93
chimpanzees, *51*, 51–53, *52*, 72–74, 90–91
chloroplasts, 194–97, *196*
choanoflagellates, 204–7, *205*
chromosomes, 61, 98–111
  activity and position of genes on, 105–11, 126–27, 155
  bundling of DNA into, 98–100, *99*, *100*, 126
  weights in different species of, 129–30
Clack, Jenny, 224–25
clavicles, 19
climbing perch, 14–15
Clinton, Bill, 75
Collins, Francis, 75
color vision, 135
comparative anatomy, 124–25
Compsognathus, 18
congenital anomalies. *See* developmental anomalies
contingency and chance, 168–92
  biological multiples and, 173–90
  evolutionary recipes and, 190–92
  Gould on cataclysm and, 168–70
  Lankester on degeneration and, 172–74, 185–86
corn, *141*, 141–43, 156

Crick, Francis, 60–62, 64
CRISPR-Cas, 113, 212–13

Darwin, Charles, xi, 30, 61, 216–17
defenders of, 5–6, 170–71
on developmental anomalies, 93–94
on embryonic development, 37
on evolutionary relationships, 67–68, 172
on gradual change, 6–9, 148–50
Haeckel's research and, 38–39
Mivart's critique of, 5–9, 14–16, 19, 22–23, 216
on repurposed function of evolving structures, 9, 15, 22–27
on variation among individuals, 183–84
See also evolution; *On the Origin of Species*
Dean, Bashford, 13–15, *14*
decidual stromal cells, 151–58, *152*
deformities. *See* developmental anomalies
degeneration, 172–74, 185–86
Deinonychus, 22–24, *23*
developmental anomalies, 54, 92–123
altered genetic activity and position in, 112–16, 126–27
chromosome mutations in, 100–111
errors in genetic duplication in, 131–34
evolution of mutations in, 151–58, 166–67
in fruit flies, 96–111, 125–27, 135
Goldschmidt's study of, 148–50
Hox genes and, 116–22, 135
in humans, 116–18
jumping genes and, 156–58
mechanisms of heredity in, 96–103, 113
developmental biology, x–xi, 13–27
linking of fossils with embryos in, 13–16, 216–18
on origins of flight, 16–27
Dexter, Stanley, 81
dinosaurs, 17–27
extinction of, 169–70
feathered coverings of, 24–27, *25*
hollow bones of, 22–24
running speed of, 21–22
semilunate bones of, 26
DNA, xi–xii, 59–91, 96, 216–17
bundling into chromosomes of, 98–100, 126
coiled inactive regions of, 89–90
evolutionary transformation and, 68–69, 87–90, 147–50

DNA *(continued)*
gene sequence in, 107–11
genetic duplication and, 139–40
genetic switches of, 77–80, 84–90, 156–58
of organelles, 195
proteins and amino acid sequences of, 62–69
sequencing of, x, 75–77, 140, 196
Sonic hedgehog gene mutations and, 80–90
structure of, 60–61
DNA technology, x–xi, 75–77, 89, 234
cutting and editing in, 113, 212–13
isolation of genes in, 107–8
Dohrn, Anton, 56
domesticated viruses, 164–65
Doudna, Jennifer, 113
Down syndrome, 132
Drosophila melanogaster. See fruit flies
duck-billed dinosaurs, 21–25
Duméril, André, 29
Duméril, Auguste, 29–31, *31*, 42–44, 174–75

Edwards syndrome, 132
Einstein, Albert, 62
embryonic development, 13–16, 28–59, 188–89, 216–17
biological diversity and, 52–53
developmental maladies in, 54
differentiation of species in, 35–37
Haeckel's recapitulation theory of, 38–41, *39*
human evolution and, 50–53
impact of timing changes on, 41–49, 54
link of fossils with, 13–16, 216–18
migration of cells during, 54–59
Naef's idealistic morphology in, 49–51
Pax genes and, 135–36
pregnancy and, 151–60
three layers of tissue in, 33–36, 41, 55–56
See also developmental anomalies
emperor penguin eggs, 39–41
Escherichia coli, 77
evolution, xi–xii, 3–27, 50–53, 215–18
biological diversity in, 52–53, 199–200, 217–18
combinations and mergers in, 194–213, *196*
contingency and chance in, 168–92
Darwin's idea of gradual change in, 6–9, 148–50
determined recipes in, 190–92
genetic duplication and, 133–45

genomic changes and transformations in, 87–90, 147–50
Haeckel's recapitulation theory of embryos and, 38–41
heredity and variation among individuals in, 93–123, 183–84
relationships among species in, 66–69, 71–74, 172, 181–84, 231–32
repurposed function of evolving structures in, 15, 22–27, 122–23, 174–81
spread of mutations in, 151–58, 166–67
See also humans
evolutionary biology, 71
Evolutionary Synthesis, 242
"Evolution at Two Levels in Humans and Chimpanzees" (King and Wilson), 90–91

fingers, 25–26
fish, 164, 193, 216
breathing of air by, 10–16
degeneration of structures in, 172–74
domesticated *Arc* gene in, 164–65
embryonic development of, 36–37
extra sets of chromosomes in, 133
genes for hands and feet in, 120–23
lungs in, *12*, 12–16, 193, 225–26
Fisher, Ronald A., 148
flight, 7–8, 16–27
biological multiples and, 182–83
links between birds and dinosaurs and, 18–27
specialized bone structures for, 25–26, 182
fossils, 13–19, 150, 198–202, 216–17
Franklin, Rosalind, 60
fruit flies *(Drosophila melanogaster)*, 96–111
Antennapedia mutant of, *107*, 107–8
Arc gene in, 165
Bithorax mutant of, *102*, 102–6, 110–11, 135
embryos of, 106
genetic duplication and, 135–36
small-eyed mutants of, 125–27

Garstang, Walter, 41–49, *42*, 54
Gassling, Mary, 81–82
Gehring, Walter, 107
Gehrke, Andrew, 121–22
Gehrke, Lee, 121
genes, 98–100
activity and position on chromosomes of, 105–11, *115*, *118*, *122*, 126–27, 155–56

genes *(continued)*
ALU sequence of, 140, 144–45
Arc gene, 162–66
building of bodies by, 112–16, 202–7
decidual stromal cells and, 151–60
homeobox genes, 240
Hox genes, *110*, 116–22, *118*, *122*, 135
jumping genes, 141–45, 156, 208, 217
LINE1 sequence of, 140
manufacture of proteins by, 60, 62–69, *79*, *85*, 108, 204
NOTCH2NL gene, 137–38
on-off switches of, 77–80, *79*, 84–90, *85*, 156–58
palindrome-space systems sequencing of, 210–12
Pax genes, 135–36
war with viruses of, 159–61, 164–67
See also DNA
genetic duplication, 124–45, 159, 217
benefits to species of, 132–33
of cancer genes, 208–9
comparative anatomy of, 124–25
errors in, 125–27, 131–32
jumping genes and, 141–45, 156–58, 208, 217
Ohno's research on, 127–34
role in evolution of, 133–45
genetics
link with Darwinian evolution of, 147–48
origin of term, 94
study of heredity and variation using, 97–123
See also evolution
genioglossus muscle, 176–80
genome browsers, 76–77, 159, 234
genome editing, 113, 212–13
genome sequencing, x, 75–77, 140, 154, 159
gliding animals, 21
Goldschmidt, Richard, 147–50, 166
Gould, Stephen Jay, 7, 51–52, 168–70, 190

Haeckel, Ernst, 38–39, *39*, 41
hair, 135
hand-foot-genital syndrome, 54
Hellman, Lillian, 4, 166
Hemingway, Ernest, 81
Hemingway cats. *See* mitten cats
hemoglobin, 134
heredity, 96–103, 113
See also developmental anomalies; fruit flies
heterochrony, 44–46
Hill, Robert, 84–86, 88
HIV (human immunodeficiency virus), 159–60, 164

homeobox genes, 240
homunculi, 32
hopeful monsters, 148–49
  See also developmental anomalies
Hox genes, *110*, 116–22, *118*, *122*, 135
Human Genome Project, x, 75–77
humans
  anatomical organization of, 125
  brain size of, 136–38
  building of bodies in, 116–18
  evolution of, 50–53, 66, 71–74, 90–91, 231–32
  genetic duplication in, 131–32
  Hox genes in, 135
  origin of bones in, 150
  pregnancy in, 151–61
  See also evolution
Huxley, Thomas Henry, 5–6, 18–19, 170–72

immunity, 211–12
individuals, 202–3
insect wings, 182–83
invertebrates, 46–49, 54
isopods, 115–16
It's a Wonderful Life (Capra), 169

Jacob, François, 77–79, 84–85
Jobs, Steve, 127
Jordan, David Starr, 56–57
jumping genes, 141–45, 156–58, 208, 217
Jurassic Age fossils, 17–19

keratin, 135
King, Mary-Claire, 71–74, 90–91
King, Nicole, 204–7
Kingsley, David, 88–89

Lankester, Ray, 170–74, *171*, 180, 182
Larval Forms and Other Verses (Garstang), 42, *42*
Lenski, Rich, 247
Leonardo da Vinci, 193
Lepidosiren paradoxa, 12–13
Lettice, Laura, 84–86
Levine, Mike, 107–9, 111
Lewis, Edward, 103–12, *104*, 114, 119, 135
limestone, 16–17
LINE1 sequence, 140, 144–45
Linnaeus, Carl, 28, 31
lizards, 190–91
Lombard, Eric, 177
Long, Manyuan, 136
Losos, Jonathan, 191
lumbar vertebrae, 117–18, *118*
lungfish, *12*, 12–13, 15, 93
Lynch, Vinny, 151–58

The Man Who Mistook His Wife for a Hat (Sacks), 161

Margulis, Lynn, 193–97, *194*, 204, 208
Marsh, O. C., 21
marsupials, 191
The Material Basis of Evolution (Goldschmidt), 147
Materials for the Study of Variation (Bateson), 94–95
Mayr, Ernst, 146–50, 166, 190, 242
McClintock, Barbara, *141*, 141–44, 151, 156, 208
McGinnis, Bill, 107–9, 111
memory, 161–66, 208
Mendel, Gregor, 94, 96
metamorphosis
  of salamanders, 43–46
  of sea squirts, 47–49
microbes, 200–207, 216
  adaptation of, 206
  metabolism of oxygen by, 201–2, 207
  organelles of, 193–97
mitochondria, 194–97, *196*
mitten cats, *80*, 80–81, 86, 96
Mivart, St. George Jackson, 5–9, *6*, 14–16, 19, 22–23, 216
Mobile Wad of Henry, 124
Mojica, Francisco, 210–12
molecular clock hypothesis, 231–32
molecular scalpel (*Cas9*), 212–13
Monod, Jacques, 77–79, 84–85
monsters, 92–95, 148–49
  See also developmental anomalies
Montagu, Ashley, 51–52
Moore, Gordon, 75
Morgan, Thomas Hunt, 95–103, 125, 131, 151
  See also fruit flies
morphometrics, 230
Mount Ritchie, 215–16, 249
mudskippers, 14–15
multiples, 173–90
Museum of Comparative Zoology (Harvard), 146
mutants. *See* developmental anomalies

Naef, Adolf, 49–51, *51*
Nakamura, Tetsuya, 121
Napoleon Bonaparte, 9–10, 93
natural selection, 183
nerve cells, 150
neurodegenerative disease, 162
Niemeyer, Jakob, 18
Nopsa von Felső-Szilvás, Franz (Baron Nopsca of Săcel), 19–21, *20*
NOTCH2NL gene, 137–38

Ohno, Susumu, 127–34, *128*, 136
On Growth and Form (Thompson), 52–53

On the Genesis of Species (Mivart), 6–8
On the Origin of Species (Darwin)
on *Archaeopteryx*, 18
first edition of, 8, 15
Huxley's defense of, 170–71
Mivart's critique of, 6–9, 14–16, 19, 22–23
on repurposed function of evolving structures, 9, 15, 22–27
sixth edition of, 8–9
on variation among individuals, 93–94
opsins, 135
organelles, 193–97, 201–2
organoids, 137–38
Ostrom, John, 21–25, *25*
Owen, Richard, 125, 127
oxygen, 201–2, 207

palindrome-space system sequences, 210–12
Pander, Christian, 33–35, 53, 55
parasitic shrimp, 172–73
Parhyale, 112–16
Patau syndrome, 132
Patel, Nipam, 112–15, 212
Pauling, Linus, 64–72, 134, 180, 231–32
Pax genes, 135–36
photosynthesis, 194, 201–2
pill bugs, 115–16
Plasmodium falciparum, 197
Plato, 50
Platt, Julia Barlow, 54–58, *57*
Pliny the Elder, 28
polydactyly, 80–87, 96
predators, 175
preeclampsia, 159
pregnancy, 151–61, 208
decidual stromal cells in, 153–60
maternal immune response to, 153
syncytin in, 158–61, 164
progesterone, 153–57
projectile tongues, 175–81, *179*
proteins, 60, 62–69, 108, 204–7
pterosaurs, 19, 169, 182
purified protein, 163

quantitative biology, 230

recapitulation theory, 38–41
Regeneration (Morgan), 96
RNA, 108
rocks, 198–202, 215–16
rules, 183

Sacks, Oliver, 161
sacral vertebrae, 117, *118*, 118–19
Saint-Hilaire, Étienne Geoffroy, 9–11, *10*, 15, 93
Saint-Hilaire, Isidore, 93
salamanders, 28–31
Duméril's two kinds of, 29–31, *31*, 43–44, 174–75

salamanders *(continued)*
embryonic skull development of, 56
evolution of biological multiples in, 180–90
evolution of limbs and feet in, 185–89, *189*
genetic duplication in, 133
metamorphosis from larvae of, 41–44
projectile tongues and gill bones of, 174–81, *179*
timing of embryonic development of, 44–46, *45*
weight of genetic material in, 130
saltwater ecosystems, 210–13
Saunders, John, 81–82
Säve-Söderbergh, Gunnar, 222–24
schizophrenia, 162
Schopf, J. William, 198–201
Scott, Matt, 109, 111
Scott, Robert Falcon, 39–40
sea squirts, 47–49, *48*, 54
semilunate bone, 26
Shepherd, Jason, 161–65
sickle cell anemia, 63
single-celled creatures, 200–201, 204–7
See also microbes
skeleton fossils, 150
skin, 135, 203–4
Slade, Henry, 171
smell receptors, 135
Sonic hedgehog gene, 80–90
sticklebacks, 88–89
swim bladders, 10–16
syncytin, 158–61, 164
Szent-Györgyi, Albert, 66

tendons, 150
theropod dinosaurs, 22–24
Thompson, D'Arcy Wentworth, *52*, 52–53
thyroid hormone, 44, 45
Tice, Sabra Cobey, 125–26
Tiktaalik roseae, ix, 164, 215, 222
Transantarctic Range, Antarctica, 215–16

Venter, Craig, 75
vertebrates, 46–49, 54–59
Vicq d'Azyr, Félix, 124–25, 127
viruses, 159–61, 164–67, 211–12, 217
Voltaire, 190
von Baer, Karl Ernst, 33–38, *34*, 41, 46, 55
von Humboldt, Alexander, 17
von Sömmerring, Samuel Thomas, 92

Wake, David, *174*, 174–82, 184–86
Watson, James, 60, 64
Whitman, O. C., 55
Wilkins, Maurice, 60
Williams, Ernest, 190–91

Wilson, Allan, 70–74, 90–91
Wonderful Life (Gould), 170
Woodworth, Charles W., 97
The Worst Journey in the World (Cherry-Garrard), 40

Your Inner Fish (Shubin), ix

zinc knuckle, 164
Zuckerkandl, Émile, 62–73, 134, 180, 231–32

ABOUT THE AUTHOR

Neil Shubin is the author of *The Universe Within* and the best-selling *Your Inner Fish*, which was chosen by the National Academy of Sciences as the best book of the year in 2009. Trained at Columbia, Harvard, and the University of California, Berkeley, Shubin is the associate dean of biological sciences at the University of Chicago. In 2011 he was elected to the National Academy of Sciences.

A NOTE ON THE TYPE

This book was set in Janson, a typeface named for the Dutchman Anton Janson but that is actually the work of Nicholas Kis (1650–1702). The type is an excellent example of the influential and sturdy Dutch types that prevailed in England up to the time William Caslon (1692–1766) developed his own incomparable designs from them.

Composed by North Market Street Graphics, Lancaster, Pennsylvania

Printed and bound by Berryville Graphics, Berryville, Virginia

Designed by Anna B. Knighton